The Geologic Story of

YOSEMITE
NATIONAL PARK

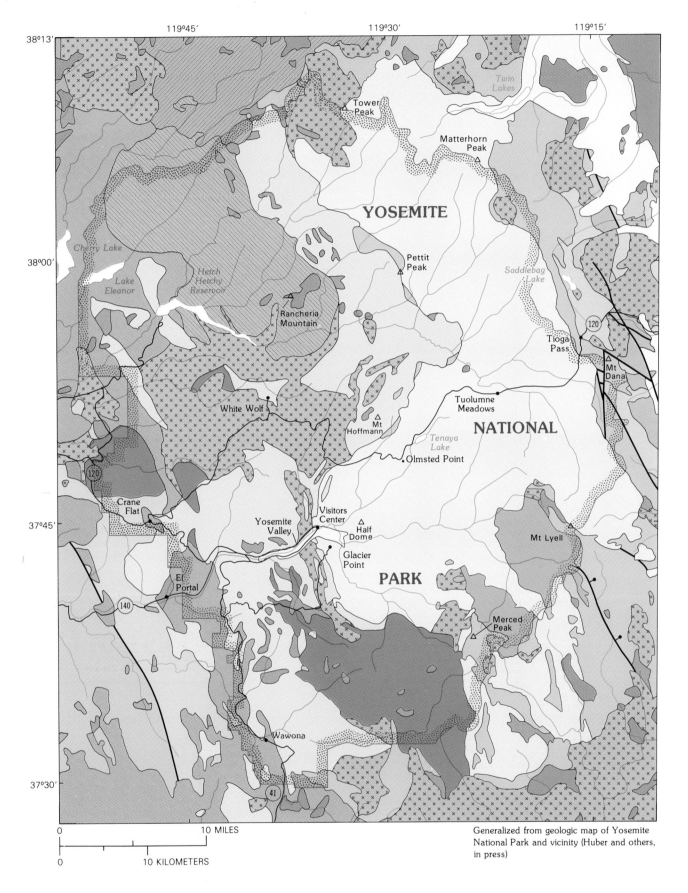

119°45' 119°30' 119°15'

38°13'
38°00'
37°45'
37°30'

YOSEMITE

NATIONAL

PARK

Twin Lakes
Tower Peak
Matterhorn Peak
Pettit Peak
Saddlebag Lake
Tioga Pass
Mt Dana
Tuolumne Meadows
Rancheria Mountain
Cherry Lake
Lake Eleanor
Hetch Hetchy Reservoir
White Wolf
Mt Hoffmann
Tenaya Lake
Olmsted Point
Crane Flat
Yosemite Valley
Visitors Center
Half Dome
Glacier Point
Mt Lyell
El Portal
Merced Peak
Wawona

120
140
41

0   10 MILES

0   10 KILOMETERS

Generalized from geologic map of Yosemite National Park and vicinity (Huber and others, in press)

GEOLOGIC MAP OF YOSEMITE NATIONAL PARK AND VICINITY

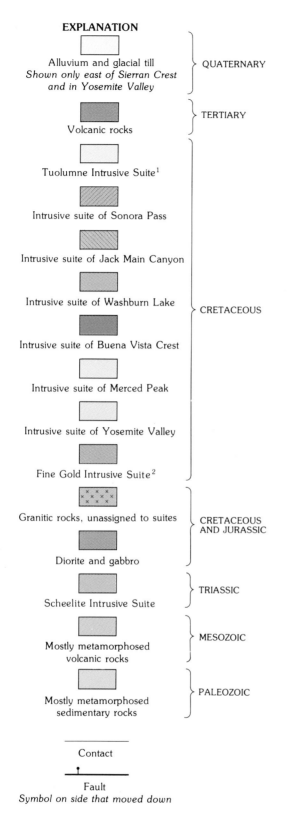

EXPLANATION

Alluvium and glacial till
*Shown only east of Sierran Crest
and in Yosemite Valley*
— QUATERNARY

Volcanic rocks
— TERTIARY

Tuolumne Intrusive Suite[1]

Intrusive suite of Sonora Pass

Intrusive suite of Jack Main Canyon

Intrusive suite of Washburn Lake

Intrusive suite of Buena Vista Crest
— CRETACEOUS

Intrusive suite of Merced Peak

Intrusive suite of Yosemite Valley

Fine Gold Intrusive Suite[2]

Granitic rocks, unassigned to suites
— CRETACEOUS AND JURASSIC

Diorite and gabbro

Scheelite Intrusive Suite
— TRIASSIC

Mostly metamorphosed
volcanic rocks
— MESOZOIC

Mostly metamorphosed
sedimentary rocks
— PALEOZOIC

———————
Contact

———•———
Fault
Symbol on side that moved down

[1] An intrusive suite is a grouping of
genetically related plutonic rocks

[2] Bateman (in press a)

GEOLOGIC MAPS OF YOSEMITE

A geologic map shows the distribution of different types of rocks at the Earth's surface. Construction of such a map is usually prerequisite to understanding the geology of any given area. Geologic quadrangle maps at the same scale as topographic quadrangle maps of the U.S. Geological Survey's 15-minute series (1 in. to 1 mi) currently are available for only a little more than half of the park area. Two other geologic maps are especially suitable to this volume because they show details of Yosemite geology not possible to show at the small scale of the geologic maps herein. A new geologic map of Yosemite National Park and vicinity, at a scale of 1 in. to about 2 mi, has recently been compiled from both published and unpublished geologic mapping by many individuals (Huber and others, in press). A more detailed geologic map of Yosemite Valley is also available (Calkins and others, 1985). The fieldwork for this latter map was carried out by Frank Calkins between 1913 and 1916, but Calkins, the consummate perfectionist, was never completely satisfied with his map and would not let it be published during his lifetime. Though now belatedly published as a historical document, it is still the best geologic map of Yosemite Valley available. These two geologic maps, reduced and generalized, are the basis for the geologic maps in this volume (pls. 1, 2).

PLATE 1

The Geologic Story of

Second Printing 1991

Yosemite Association
Box 545
Yosemite National Park, California 95389

ISBN-0-939666-49-9

**Previously published as
"U.S. Geological Survey Bulletin 1595"**

YOSEMITE
NATIONAL PARK

A comprehensive geologic view of the natural processes that have created— and are still creating— the stunning terrain we know as Yosemite

By N. King Huber

YOSEMITE ASSOCIATION
YOSEMITE NATIONAL PARK, CALIFORNIA
1989

HALF DOME AT SUNSET

*"On the south-east stands the majestic Mount Tis-sa-ack, or 'South Dome' * * *. Almost one-half of this immense mass, either from some convulsion of nature, or 'Time's effacing fingers,' has fallen over * * *. Yet proudly, aye, defiantly erect, it still holds its noble head, and is not only the highest of all those around, but is the greatest attraction of the valley."*

—J. M. Hutchings, "Scenes of Wonder and Curiosity in California," 1870.

FOREWORD

Within 150 years, Yosemite has moved from great obscurity to worldwide fame as one of the most visited of our national parks. As a remarkable place where people can enjoy unparalleled scenes of natural beauty and where many easily observed geologic features are concentrated, the park is rivaled by few other areas on the planet. The majesty and immense variety of these features have inspired artists and photographers, intrigued tourists, and stirred controversy among geologists.

Field studies in the Yosemite area have contributed to the development of our ideas about geologic processes, including the different actions of streams and glaciers in the evolution of the landscape, and the formation of granite, the basic bedrock of much of the Earth's continents. The park's role as a natural laboratory for geologic research cannot be overemphasized, and its investigation has led to many landmark studies by U.S. Geological Survey geologists over the past 70 years. In 1913, the first detailed program of research on the geology of the park and the origin of Yosemite Valley was begun by François Matthes and Frank Calkins. Their work, along with that of later generations of Survey geologists, myself included, serves as the basis for our present understanding of the geologic history of Yosemite and of the processes that formed and continue to mold its landscape.

This book, which makes available in one volume a comprehensive summary of the current geologic knowledge of Yosemite National Park, is an excellent example of the Survey's continuing effort to provide earth-science information in the public service.

Dallas L. Peck
Director, U.S. Geological Survey

CONTENTS

::

Special topics ::

ILLUSTRATIONS

The Geologic Story of
YOSEMITE
NATIONAL PARK

By N. King Huber

YOSEMITE COUNTRY

CREATION OF A PARK

For its towering cliffs, spectacular waterfalls, granite domes and spires, glacially polished rock, and groves of Big Trees, Yosemite is world famous. Nowhere else are all these exceptional features so well displayed and so easily accessible. Artists, writers, tourists, and geologists have flocked to Yosemite—and marveled.

Although there are other valleys with similarities to Yosemite, there is but one Yosemite Valley, the "Incomparable Valley" of John Muir. Appreciation of Yosemite Valley came early, and in 1864, less than 15 years after the general public became aware of the area's existence, President Abraham Lincoln signed a bill that granted Yosemite Valley—"the 'Cleft' or 'Gorge' in the granite peak of the Sierra Nevada"—to the State of California. The act stipulated that "the premises shall be held for public use, resort, and recreation; shall be inalienable for all time." Also included in the grant was the "Mariposa Big Tree Grove." Though not the first official national park, Yosemite established the national-park concept and eventually evolved into a national park itself. An area larger than the present park, surrounding but not including Yosemite Valley, was set aside as a national park in 1890. In 1906, the boundaries were adjusted, and Yosemite Valley and the Mariposa Grove were re-ceded to the Federal Government by California to create a unified national park (fig. 1).

YEARS OF EXPLORATION

From the earliest days, the Sierra Nevada (Spanish for "snowy mountain range") was a formidable barrier to westward exploration (fig. 2). Running half the length of California, it is the longest, the highest, and the grandest continuous mountain range in the United

YOSEMITE NATIONAL PARK and the original grants to the State of California. (Fig. 1)

THE SIERRA NEVADA, a strongly asymmetric mountain range with a steep east escarpment and a gentle westward slope toward the broad Central Valley of California. Physiography from landform map by Erwin Raisz; used with permission. (Fig. 2)

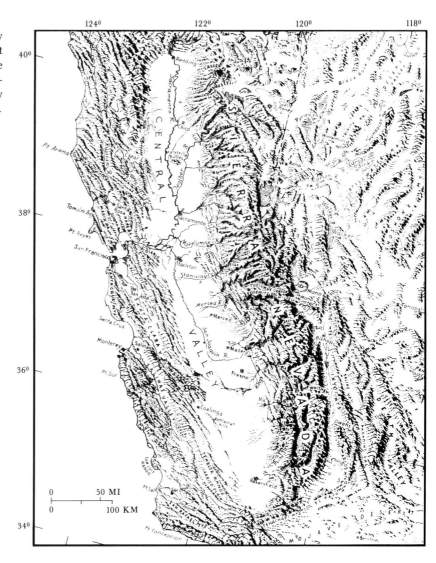

States, outside of Alaska. The central Sierra with its steep east escarpment is particularly awesome. Nevertheless, in 1833, Joseph Walker led a party up the east escarpment and westward across the range through Yosemite country. His route traversed the upland between the Tuolumne and Merced Rivers, a route later followed by the western part of the Tioga Road. Walker and his men were probably the first of European descent to view Yosemite Valley and the Big Trees, now known as giant sequoias.

The Walker party's journal, recorded by Zenas Leonard, refers to "many small streams which would shoot out from under these high snow-banks, and after running a short distance in deep chasms which they have through ages cut in the rocks, precipitate themselves from one lofty precipice to another, until they are exhausted in rain below. Some of these precipices appear to be more than a mile high. * * * we found it utterly impossible to descend, to say nothing of the

horses. [Continuing westward] * * * we have found some trees of the redwood species, incredibly large— some of which would measure from 16 to 18 fathoms [96 to 108 ft] around the trunk at the height of a man's head from the ground." The trees Leonard described could be those either of the Tuolumne Grove or of the Merced Grove, possibly both. The journal was printed in Pennsylvania in 1839, but only a few copies survived a printshop fire, and so this account went unread for many years.

Yosemite Valley and the giant sequoias remained unknown to the world at large for nearly another 20 years after the Walker party's discovery, until Maj. James Savage and the Mariposa Battalion of militia entered the valley in pursuit of Indians in 1851. Overwhelmed by the majesty of the valley, one member of the battalion, Dr. Lafayette Bunnell, remarked that it needed an appropriate name. He suggested Yo-sem-i-ty, the name of the Indian tribe that inhabited it, and also

SUMMIT OF MOUNT HOFFMANN. Charles F. Hoffmann, cartographer with the Geological Survey of California, at the transit. Photograph by W. Harris, 1867; first published in J.D. Whitney's "The Yosemite Book" in 1868. (Fig. 3)

the Indian word for grizzly bear. A year later, giant sequoias were discovered anew in the Mariposa Grove and in the Calaveras Grove north of Yosemite.

The history of further exploration of the Yosemite area, and of the creation of the park itself, were well described by Carl P. Russell (1957). Of particular geologic interest was the excursion of the Geological Survey of California to Yosemite in 1863. After visiting Yosemite Valley, Josiah Whitney, the Director of the Survey, accompanied by William Brewer and Charles Hoffmann, explored the headwaters of the Tuolumne River and named Mounts Dana, Lyell, and Maclure for famous geologists and Mount Hoffmann for one of their own party. In 1867, another party from the Geological Survey of California again ascended Mount Hoffmann, accompanied by photographer W. Harris, who documented the scene with Hoffmann himself at the transit (fig. 3). Observations from these excursions, and additional topographic mapping by Geological Survey of California colleagues Clarence King and James Gardiner, provided the first description of Yosemite Valley and the High Sierra that not only contained reasonably accurate topographic information but also was relatively free from the romantic exaggeration characteristic of the times. The term "High Sierra," coined by Whitney to include the higher region of the Sierra Nevada, much of it above timberline, has been used by writers and hikers ever since. Whitney and his party recognized abundant evidence for past glaciation in the High Sierra but failed to recognize the degree to which glaciers had modified the topography, and Whitney

ascribed the origin of Yosemite Valley to a "grand cataclysm" in which the bottom simply dropped down.

Indeed, most geologic processes were poorly understood in Whitney's day, and so numerous conflicting interpretations soon developed regarding the origin of many of Yosemite's scenic features. The controversy that arose between Josiah Whitney and John Muir regarding the origin of Yosemite Valley reflects this situation. Muir's observations in the Yosemite Sierra led him to propose that Yosemite Valley was entirely carved by a glacier. However, he overestimated both the work of glaciers and the extent of glaciation, because he believed that ice once completely covered the Sierra to the Central Valley and beyond. Thus, Whitney and Muir held opposing views that were both too extreme, although Muir's ultimately proved more durable. Finally, partly in response to this controversy, a study of the geology of the Yosemite area was initiated in 1913 by the U.S. Geological Survey, with François E. Matthes studying the geomorphology and glacial geology and Frank C. Calkins the bedrock geology. Matthes' conclusions, particularly with respect to the relative roles of rivers and glaciers in sculpting the landscape, have held up well, and his lucid descriptions and interpretations have enlightened many a park visitor. In the 50-odd years since Matthes and Calkins completed their studies, we have gained considerably more geologic knowledge of the Sierra Nevada; we have abandoned some of their ideas, but we still build on their pioneering efforts.

3

VIEWS FROM MOUNT HOFFMANN. *A*, Westerly view down wooded slopes toward California's Central Valley. *B*, Northerly view including snow-patched Sawtooth Ridge and Matterhorn Peak on Skyline at north edge of the park. Photograph by Tau Rho Alpha. *C*, Easterly view, with May Lake in foreground and Tenaya Lake to right in middle distance. Tuolumne Meadows is in wooded area to left, and Mount Dana is highest summit on skyline beyond. *D*, Southerly view toward Clouds Rest in late spring, with Mount Clark on left skyline. Half Dome at far right center displays its northeast shoulder, in contrast to its oft-pictured profile from Yosemite Valley's floor. Photograph from National Park Service collection. (Fig. 4)

A

B

C

D

A VIEW FROM THE TOP—MOUNT HOFFMANN

Of Mount Hoffmann, William Brewer noted in his journal under the date June 24, 1863, that "It commanded a sublime view and the scene is one to be remembered for a lifetime." In his turn, Josiah Whitney stated that "The view from the summit of Mount Hoffmann is remarkably fine."

The view from Mount Hoffmann is, indeed, remarkably fine. This peak is in almost the exact center of Yosemite Park (fig. 4), and from its 10,850-ft summit we can see much of the perimeter of the park, for much of that perimeter consists of high ridges.

Looking westward from Mount Hoffmann, we see timbered foothills disappearing into the haze of the Central Valley of California. As we shift our view northward, we see the smooth contours of the foothills give way along the skyline to pointed and jagged peaks of bare rock in shades of white, red, and gray. A lake-strewn cirque, which fed the Hoffmann Glacier that moved down Yosemite Creek, forms the precipitous north face of Mount Hoffmann itself.

To the east, May Lake—named for Hoffmann's wife—is directly below and Tenaya Lake is in the middle distance, surrounded by sheeted granite walls,

which lead to a series of monolithic granite domes between Tenaya Lake and Tuolumne Meadows—domes that are among the most striking features of the Yosemite landscape. On the eastern segment of the skyline rise the highest peaks in the park: Conness, Dana, Gibbs, Koip, and Lyell—all but Gibbs with living glaciers. The vista to the south past Clouds Rest, Half Dome, and Yosemite Valley has the Clark Range and Buena Vista Crest as the skyline backdrop. From Mount Hoffmann we see the northeast shoulder of Half Dome, a striking contrast to the view from Yosemite Valley (figs. 5, 6).

The scenic panorama from Mount Hoffmann is a grand introduction to much of the geology of Yosemite. The landforms of Yosemite, like landforms everywhere, reflect the type, structure, and erosional history of the underlying rocks. The different colors—white, red, shades of gray—reflect different rock types. The different topographic shapes—spires, domes, cliffs—reflect different rock structures and erosional histories, especially erosion caused by glaciers. In discussing "how best to spend one's Yosemite time," John Muir suggested "go straight to Mount Hoffmann. From the summit nearly all the Yosemite Park is displayed like a map." For those who wish to take his advice, the summit is about 3 mi from the trailhead south of May Lake, and the elevation gain is about 2,000 ft.

The landscape we see today is largely the result of geologic processes operating in the past few tens of millions of years on the parent rock. But to gain a full understanding of the present landscape, we must go back many tens of millions of years more—to the creation of the rocks themselves. Fragments of geologic history going back hundreds of millions of years can be read from the rocks of the park, and with additional data from elsewhere in the Sierra Nevada and beyond, we can reconstruct much of the geologic story of Yosemite. After more than a hundred years of study, the story is still incomplete. But then, geologic stories are seldom complete, and what we do know whets our curiosity about the missing pieces and allows a deeper appreciation for one of our most spectacular national parks. This volume is an attempt to describe the geology of Yosemite and to explain how this splendid landscape, centered on Mount Hoffmann, came into being.

PANORAMA FROM MOUNT HOFFMANN, encompassing an easterly-facing arc between Whorl Mountain on left and Half Dome on right. (Fig. 5)

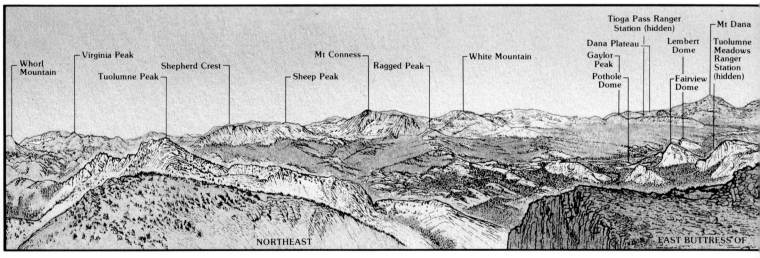

GEOLOGIC OVERVIEW

Topographically, the Sierra Nevada is an asymmetric mountain range with a long, gentle west slope and a short, steep east escarpment that culminates in the highest peaks (fig. 2). It is 50 to 80 mi wide and extends in altitude from near sea level along its west edge to more than 13,000 ft along the crest in the Yosemite area. Geologically, the Sierra Nevada is a huge block of the Earth's crust that has broken free on the east along a bounding fault system and has been uplifted and tilted westward. This combination of uplift and tilt, which is the underlying geologic process that created the present range, is still going on today.

Massive granite dominates the Yosemite area and much of the Sierra Nevada as well. Mount Hoffmann and most of the terrane visible from it are composed of granite, formed deep within the Earth by solidification of formerly molten rock material and subsequently exposed by erosion of the overlying rocks. Because of its massiveness and durability, granite is shaped into bold forms: the cliffs of Yosemite and Hetch Hetchy Valleys,

many of the higher peaks in the park, and the striking sheeted domes that can form only in massive, unlayered rock. Although granite dominates nearly the entire length of the Sierra, the granite is not monolithic. Instead, it is a composite of hundreds of smaller bodies of granitic rock that, as magma (molten material), individually intruded one another over a timespan of more than 100 million years (fig. 7). This multiplicity of intrusions is one of the reasons why there are so many varieties of granitic rock in Yosemite and the rest of the Sierra. The differences are not always apparent to the casual observer, but they are reflected in sometimes subtle differences in appearance and in differences in response to weathering and erosion acting on the rocks.

Layered metamorphic rocks in the foothills at the west edge of the park and along the eastern margin in the summit area are remnants of ancient sedimentary and volcanic rocks that were deformed and metamorphosed in part by the invading granitic intrusions. Other metamorphic rocks that once formed the roof beneath which the granitic rocks solidified were long ago eroded away to expose the granitic core of the range, and only small isolated remnants are left.

COMMON MINERALS IN GRANITE

Five minerals compose the bulk of the plutonic rocks of Yosemite: quartz, potassium feldspar, plagioclase feldspar, biotite, and hornblende. Quartz and both varieties of feldspar are translucent and appear light gray on fresh surfaces. On a weathered surface, the feldspars turn chalky white, whereas the quartz remains clear gray. Feldspar crystals have good cleavage, a property of breaking along planar surfaces that reflect sunlight when properly oriented; quartz has no cleavage but breaks randomly along curved surfaces. Biotite crystals commonly appear hexagonal, and their dark, brown to black plates can be split with a knife into thin flakes along one perfect cleavage direction (fig. 8). Hornblende is much harder than biotite, appears very dark green to almost black, and commonly occurs as elongate, rod-shaped crystals. It has good cleavages in two directions that intersect to form fine striations along the length of the rods, making them look like bits of charcoal. Other minerals are present in small amounts; the most distinctive is sphene (calcium and titanium silicate), which occurs in small, amber, wedge-shaped crystals. With a little practice, all these minerals can be identified with a small magnifying glass.

HORNBLENDE AND BIOTITE. Rod-shaped crystals of hornblende and hexagonal crystals of biotite. These large and exceptionally well formed crystals are from Half Dome Granodiorite. (Fig. 8)

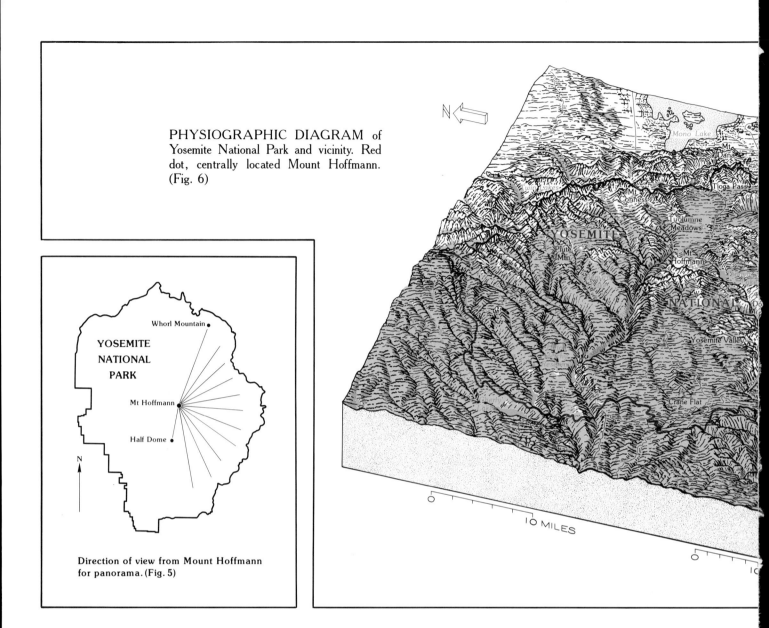

PHYSIOGRAPHIC DIAGRAM of Yosemite National Park and vicinity. Red dot, centrally located Mount Hoffmann. (Fig. 6)

YOSEMITE NATIONAL PARK

Whorl Mountain

Mt Hoffmann

Half Dome

N

Direction of view from Mount Hoffmann for panorama. (Fig. 5)

N

Mono Lake

Mt Dana

Tioga Pass

Mt Conness

Tuolumne Meadows

YOSEMITE

Piute Mtn

Mt Hoffmann

NATIONAL PARK

Yosemite Valley

Crane Flat

0 [ ] 10 MILES

0

10

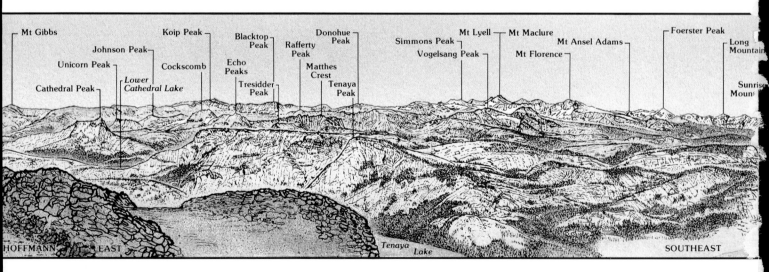

Mt Gibbs

Johnson Peak

Unicorn Peak

Cathedral Peak

Koip Peak

Cockscomb

Lower Cathedral Lake

Blacktop Peak

Echo Peaks

Tresidder Peak

Rafferty Peak

Matthes Crest

Donohue Peak

Tenaya Peak

Simmons Peak

Vogelsang Peak

Mt Lyell

Mt Florence

Mt Maclure

Mt Ansel Adams

Foerster Peak

Long Mountain

Sunrise Mount

HOFFMANN   EAST

Tenaya Lake

SOUTHEAST

7

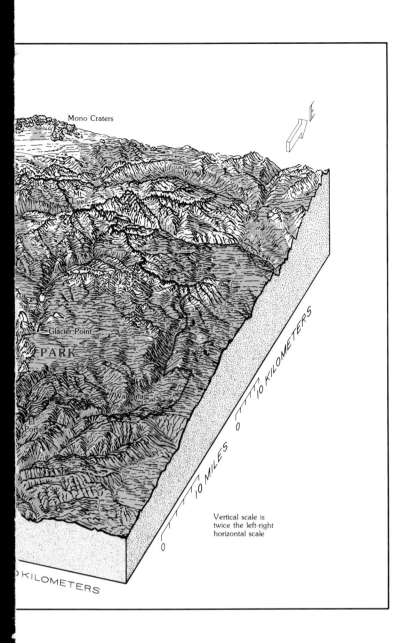

Mono Craters

Mt
Lyell

Glacier Point

PARK

Wawona

El Portal

10 MILES

10 KILOMETERS

10 KILOMETERS

Vertical scale is
twice the left-right
horizontal scale

KILOMETERS

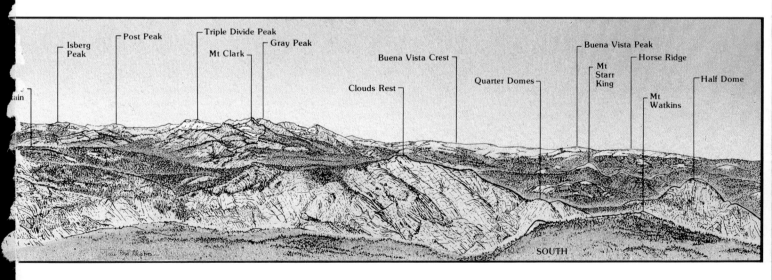

Isberg
Peak

Post Peak

Triple Divide Peak

Mt Clark

Gray Peak

Buena Vista Crest

Clouds Rest

Quarter Domes

Buena Vista Peak

Mt
Starr
King

Horse Ridge

Mt
Watkins

Half Dome

SOUTH

8

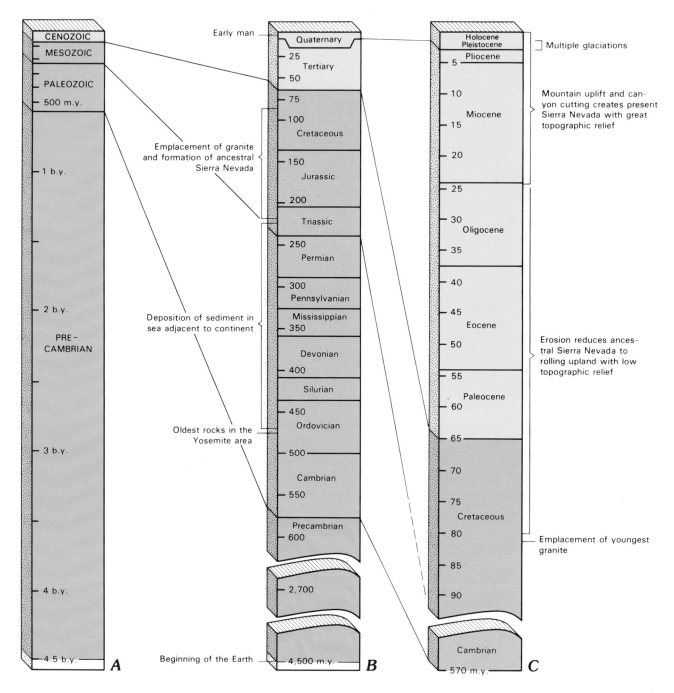

Column A

CENOZOIC
MESOZOIC
PALEOZOIC
— 500 m.y.
— 1 b.y.
— 2 b.y.
PRE-CAMBRIAN
— 3 b.y.
— 4 b.y.
— 4.5 b.y.   **A**

Column B

Early man — Quaternary
— 25   Tertiary
— 50
— 75
— 100   Cretaceous
— 150   Jurassic
— 200
Triassic
— 250   Permian
— 300   Pennsylvanian
Mississippian
— 350
Devonian
— 400
Silurian
— 450   Ordovician
— 500
Cambrian
— 550
Precambrian
— 600
— 2,700
Beginning of the Earth — 4,500 m.y.   **B**

Emplacement of granite and formation of ancestral Sierra Nevada

Deposition of sediment in sea adjacent to continent

Oldest rocks in the Yosemite area

Column C

Holocene
Pleistocene   □ Multiple glaciations
— 5   Pliocene
— 10
Miocene
— 15
— 20
— 25
— 30   Oligocene
— 35
— 40
— 45   Eocene
— 50
— 55
Paleocene
— 60
— 65
— 70
— 75   Cretaceous
— 80
— 85
— 90
Cambrian
— 570 m.y.   **C**

Mountain uplift and canyon cutting creates present Sierra Nevada with great topographic relief

Erosion reduces ancestral Sierra Nevada to rolling upland with low topographic relief

Emplacement of youngest granite

THE GEOLOGIC TIME SCALE—the "calendar" used by geologists in interpreting Earth history. Column A, graduated in billions of years (b.y.) and subdivided into the four major geologic eras (Precambrian, for example), represents the time elapsed since the beginning of the Earth, which is believed to have been about 4.5 b.y. ago. Column B is an expansion of part of the time scale in millions of years (m.y.), to show the subdivisions (periods—Cambrian, for example) of the Paleozoic, Mesozoic, and Cenozoic Eras; column C is a further expansion to show particularly the subdivisions (epochs—Paleocene, for example) of the Tertiary and Quaternary Periods. Some key events in the geologic history of Yosemite National Park are listed alongside the columns, opposite the time intervals in which they occurred.

The subdivisions of geologic time are based largely on the fossil record; rocks of the Cambrian Period contain the earliest evidence of complex forms of life, which evolved through subsequent periods into the life of the modern world. The ages (in years) are based on radiometric dating. Many rocks contain radioactive elements that begin to decay at a very slow but measurable rate as soon as the parent rock is formed. The most common radioactive elements are uranium, rubidium, and potassium, and their decay ("daughter") products are lead, strontium, and argon, respectively. By measuring both the amount of a given daughter product and the amount of the original radioactive element still remaining in the parent rock, and then relating these measurements to the known rates of radioactive decay, the age of the rock in actual numbers of years can be calculated. (Fig. 7)

Because Yosemite is centered on this deeply dissected body of granite, metamorphic rocks are sparse; they occupy less than 5 percent of the area of the park.

Evolution of the landscape is as much a part of the geologic story as the rocks themselves, and Yosemite is a place where the dynamism of geologic processes is well displayed. By the end of Cretaceous time (see fig. 7), about 65 million years ago, after the granite core of the range had been exposed, the area had a low relief in comparison with the mountains of today. Then, about 25 million years ago, this lowland area began to be uplifted and tilted toward the southwest, a construction that would eventually lead to the present Sierra Nevada. As the rate and degree of southwestward tilt increased, the gradients of streams flowing southwestward to California's Central Valley also increased, and the faster flowing streams cut deeper and deeper canyons into the mountain block. About 10 million years ago, from the Tuolumne River northward, these canyons were inundated and buried by volcanic lava flows and mudflows, and the streams were forced to begin their downcutting anew, in many places shifting laterally to find a new route to the Central Valley. The streams were equal to the task, however, and the present river courses and drainage patterns throughout the Sierra became well established.

As the world grew colder, beginning about 2 or 3 million years ago, the Sierra Nevada had risen high enough for glaciers and a mountain icefield to form periodically along the range crest. When extensive, the icefield covered much of the higher Yosemite area and sent glaciers down many of the valleys. Glacial ice quarried loose and transported vast volumes of rubble, and used it to help scour and modify the landscape. Much of this debris eventually accumulated along the margins of the glaciers and in widely distributed, hummocky piles. The greatest bulk of this debris, however, was flushed out of the Sierra to the Central Valley by streams swollen with meltwater formerly stored in the glaciers as ice and released as the glaciers melted away.

Although many of today's general landforms existed before modification by glacial action, some of them surely did not. Can you imagine the Yosemite landscape with no lakes? Virtually all the innumerable natural lakes in the park are the result of glacial activity. But even these lakes are transitory, doomed to be filled with sediment and become meadows; many lakes already have undergone this transformation. Yosemite Valley itself once contained a lake.

The geologic story of Yosemite National Park can be considered in two parts: (1) deposition and deformation of the metamorphic rocks and emplacement of the granitic rocks during the Paleozoic and Mesozoic; and (2) later uplift, erosion, and glaciation of the rocks during the Cenozoic to form today's landscape.

The paragraphs that follow start with a description of the rocks—what can be seen on excursions through the park—granite first and in the most detail, because it dominates the Yosemite scene. The rocks will then be fitted into the context of a geologic history through which today's Yosemite evolved.

ROCKS, THE BUILDING MATERIALS

Yosemite is renowned for its magnificent rock exposures. Although granitic rocks dominate the Yosemite scene, various metamorphic and volcanic rocks are also present. Together, these rocks form Yosemite's foundation.

GRANITE, GRANITE EVERYWHERE

Granite, in the broad sense of the term, is a massive rock with a salt-and-pepper appearance due to random distribution of light and dark minerals. The mineral grains are coarse enough to be individually visible to the naked eye.

Granite is a plutonic igneous rock. There are two types of igneous rock—plutonic and volcanic. Both types result from the cooling and solidification of molten rock, or *magma*. Magma originates deep within the Earth and rises toward the Earth's surface at temperatures of about 1,000 °C if granitic in composition and of as high as 1,200 °C if basaltic—by comparison, steel melts at about 1,430 °C. Magma that cools and solidifies within the Earth's crust forms plutonic rock (named for Pluto, the Roman god of the underworld). The slow cooling of plutonic magma fosters the growth of individual crystals visible to the naked eye. In contrast, magma that erupts at the Earth's surface, where it is known as lava, quickly cools into volcanic rock. Thus, having insufficient time to grow, most mineral grains in volcanic rock are so small that a microscope is needed to distinguish them.

The plutonic terrane in the Sierra, once thought simply to represent local variations in one huge mass

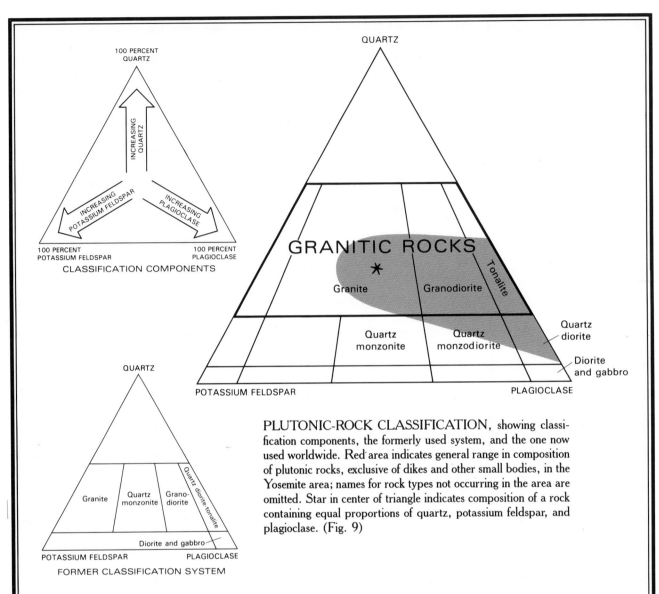

PLUTONIC-ROCK CLASSIFICATION, showing classification components, the formerly used system, and the one now used worldwide. Red area indicates general range in composition of plutonic rocks, exclusive of dikes and other small bodies, in the Yosemite area; names for rock types not occurring in the area are omitted. Star in center of triangle indicates composition of a rock containing equal proportions of quartz, potassium feldspar, and plagioclase. (Fig. 9)

—CLASSIFICATION OF PLUTONIC ROCKS—

Names for the more common varieties of plutonic rocks are based on the relative proportions of quartz, potassium feldspar, and plagioclase, as plotted on a triangular diagram, with each corner representing 100 percent of that constituent (fig. 9); other minerals present are ignored. The greater the percentage of any one of these three minerals in the rock, the closer the rock's composition would plot to the corner for that mineral. A rock with equal percentages of the three minerals would plot in the center of the diagram (*), and the rock would be called a granite. Increasing the percentage of plagioclase at the expense of potassium feldspar would move the composition toward the granodiorite compartment on the triangular diagram. "Granitic rocks" are those that lie within the heavy-lined boundary.

The rock classification used in this volume was adopted by an international commission in 1972 and is now used worldwide. This classification differs from the one previously in use and thus results in many contradictions with the rock names in earlier geologic writings on the Sierra Nevada. Nearly all the granitic rocks in the Sierra previously called quartz monzonite fall within the granite classification of the present system, and quartz monzonite is relegated to a small compartment below granite on the triangular diagram; the old system is shown for comparison. In some cases, rocks previously called quartz monzonite are now called granodiorite because of better knowledge of their actual mineral composition.

of granite, is actually made up of many individual bodies of plutonic rock—*plutons*—that formed from repeated intrusions of magma into older host rocks beneath the surface of the Earth. These plutonic rocks, formerly deep within the Earth, are now exposed at the surface, owing to deep erosion and removal of the formerly overlying rocks; they form the monoliths and domes of Yosemite within the lofty Sierra Nevada.

The collection of plutons in the park is part of a larger mass of plutonic rock called the Sierra Nevada batholith (from the Greek words *bathos*, deep, and *lithos*, rock). Although this large mass of granite forms the bedrock of much of the Sierra Nevada, it is different from the range itself and originated many tens of millions of years before uplift, weathering, and erosion shaped the present range. It needs to be emphasized that the batholith is composite, a fact not perceived by the earliest geologic studies. Distinguishing between individual plutons that represent separate episodes of intrusion and solidification of magma is the key to understanding the origin and complex geologic history of the batholith. Geologists have mapped more than a hundred discrete masses of plutonic rock in the vicinity of Yosemite National Park alone, attesting to the complexity of what was once thought to be a relatively simple batholithic setting. Emplacement of the Sierra Nevada batholith at depth may have taken as long as 130 million years.

Five minerals compose the bulk of the plutonic rocks of the batholith: quartz, two varieties of feldspar (potassium feldspar and plagioclase), biotite, and hornblende. All contain the elements silicon and oxygen, and all except quartz contain aluminum as well. Other constituents of the feldspars include potassium, sodium, and calcium; greenish-black hornblende and the black mica, biotite, also contain magnesium and iron. The section on common minerals in granite provides clues on how to identify these minerals.

Plutonic rocks consisting chiefly of quartz and feldspar, with only a minor amount of dark minerals, are loosely called granitic rocks. Granitic rocks, such as granite, granodiorite, and tonalite, differ primarily in the relative proportions of these minerals (fig. 9). For example, granite, in the technical sense of the term, contains much quartz and both potassium feldspar and calcium-rich feldspar (plagioclase). In outcrop, it is generally difficult to distinguish the relative percentages of potassium feldspar and plagioclase. In the laboratory, the feldspars can be distinguished by applying chemicals that stain potassium feldspar yellow, plagioclase red, and leave quartz uncolored (fig. 10). By this means, the relative percentages of the three minerals can be determined easily.

Granodiorite (fig. 11) is similar to granite but contains about twice as much plagioclase as potassium feldspar. Tonalite contains even less potassium feldspar. In addition to quartz and feldspar, dark minerals, such as hornblende and biotite, further characterize individual plutonic-rock types, as is commonly indicated with modified names, such as hornblende granodiorite and biotite granodiorite. Dark minerals are generally more abundant where potassium feldspar is scarce, and thus granodiorite tends to be darker than granite, and most tonalite even darker.

A

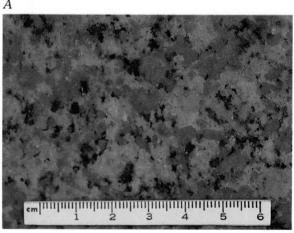

B

EL CAPITAN GRANITE. *A*, Freshly broken surface of the rock. *B*, Rock surface chemically etched and stained to differentiate potassium feldspar (orange yellow), plagioclase (red), and quartz (uncolored). (Fig. 10)

13

A

B

C

D

VARIETIES OF GRANODIORITE. All these granodio-rites have about the same mineral composition but differ in texture: Half Dome Granodiorite (*A*) contains large, well-formed hornblende crystals; Sentinel Granodiorite (*B*) contains both biotite and hornblende in poorly formed crystals; Leaning Tower Granodiorite (*C*) has a spotted appearance from rounded clots of dark minerals; and Bridalveil Granodiorite (*D*) has a salt-and-pepper appearance from fine, evenly distributed light and dark minerals. (Fig. 11)

DIORITE is mostly plagioclase and dark minerals, with little quartz and potassium feldspar. (Fig. 12) _____

In contrast to granitic rocks, quartz diorite, diorite, and gabbro contain mostly plagioclase and dark miner-als, with little or no quartz or potassium feldspar (figs. 9, 12). In addition, the plagioclase in gabbro contains more calcium than the plagioclase in diorite. Such plutonic rocks poor in quartz are sparse in the Yosemite area and generally occur as small, irregular masses and dikes—sheetlike masses—of quartz diorite or diorite; they generally are dark gray and commonly are fine grained, with few minerals readily recognizable to the naked eye.

Light-colored rock, composed chiefly of quartz and potassium feldspar, also forms irregular masses and dikes. This rock occurs both with a fine-grained tex-ture—*aplite* (fig. 13)—and with a very coarse grained texture—*pegmatite*—displaying large, intergrown quartz and potassium feldspar crystals. A fine example of pegmatite is visible a short distance down the Pohono Trail to Taft Point from the Glacier Point Road.

Most granitic rocks contain mineral grains of about equal size and are said to have a granular texture. Some granites, however, and many volcanic rocks have crystals of one mineral considerably larger than the others; these oversized crystals are called *phenocrysts* (from the Greek words meaning "to appear" and "crystal"), and the texture of such a rock is described as *porphyritic*. In Sierran granites, the most common mineral to occur as phenocrysts is potassium feldspar, in crystals commonly as much as 2 to 3 in. long (fig. 14).

Rounded inclusions of dark, fine-grained dioritic material are common in granitic rocks, most commonly in granodiorites and tonalites. Generally pancake or football shaped, the inclusions range in size from a few inches to many feet across. It is not uncommon for all the inclusions within an area to have their long dimensions arranged in the same direction, like a school of fish (fig. 15). An excellent example occurs at the Yosemite Falls overlook on the north rim of the valley. The origin of these inclusions is uncertain. Some probably were derived from preexisting rock; others may be derived from globules of darker magma that because of their high melting temperature were chilled by the granitic magma rather than being digested into it. However, the shape of the inclusions suggests that, whatever their origin, they were at least partially plastic while suspended in the magma and that they were stretched and given their parallelism by movement within the magma.

Concentrations of dark minerals sometimes form wavy, discontinuous streaks and layers, especially near the outer margins of individual plutons. These layers, called *schlieren* (German for streaks), probably represent clustering of dark minerals early during the crystallization of the magma, with alignment in streaks caused by movement within the partially solidified magma (fig. 16). The commonly abrupt termination of one set of layers by another set suggests repeated pulses of movement in a magma mush.

Individual bodies of granitic rock, particularly large ones, generally vary in mineral makeup and commonly overlap the boundaries between specific rock classifications. Bodies of granitic rock may also overlap each other's compositional ranges, and so composition is only one factor in the recognition of separate rock

DIKE of light-colored, fine-grained aplite crosscutting granodiorite. Aplite is a silica-rich rock composed chiefly of quartz and potassium feldspar. (Fig. 13)

PORPHYRITIC TEXTURE in Cathedral Peak Granodiorite, with potassium feldspar phenocrysts much larger than the other minerals in the rock matrix. (Fig. 14)

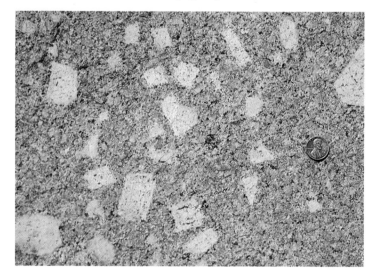

ALIGNED DARK DIORITIC INCLUSIONS in granodiorite. Photograph by Dallas L. Peck. (Fig. 15)

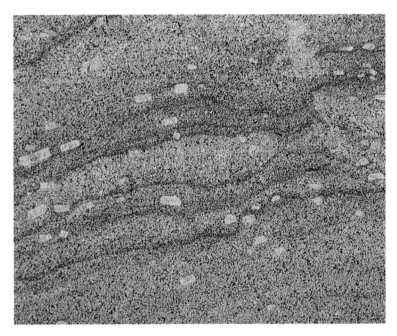

SCHLIEREN—streaks or layers formed by clustering of dark minerals during differential flow within the partially solidified magma. Note parallel alignment of potassium feldspar phenocrysts by the same process; larger phenocrysts are about 2 in. long. (Fig. 16)

bodies. The chief distinguishing property may be the presence or absence of specific minerals, such as biotite, hornblende, or sphene. Or it may be the general physical appearance defined by the texture of the rock—the size, shape, and arrangement (random or oriented) of the minerals (fig. 11). A porphyritic texture is particularly useful because it is prevalent in only a few plutons in the Yosemite area. The presence or absence of dark inclusions may also characterize a rock body.

Knowledge of the age relations among plutons is essential to understanding the geologic history of the Sierra Nevada batholith. Certain features observed in outcrop help determine the relative ages of individual rock bodies. For example, younger magma commonly shoots thin sheets, or dikes, into cracks in the older rocks (fig. 17A). Additionally, some of the younger plutons contain inclusions, or fragments of older rock, which were embedded in the younger rock while it was still molten (fig. 17B). Where dark inclusions or other oriented structures are present, the contact between two rock bodies may truncate structures in the older body, while similar features in the younger rock may parallel the contact (fig. 17C).

Determining the absolute age of a given granitic rock, in millions of years, requires measurement of the extent of radioactive decay of certain elements, such as uranium, potassium, and rubidium. From such measurements and the known rates of decay, we can approximately determine the time elapsed since the rock crystallized or cooled enough to stop escape of the daughter decay products from the rock (see fig. 7).

In their studies of plutonic rocks, geologists have devised ways to separate individual bodies of such rock and to depict them on geologic maps so as to show their relations to each other and to nonplutonic rocks with which they are in contact. Once established by field study, the boundaries of these individual plutonic-rock bodies—plutons—can be plotted on a map, and these rock bodies become geologic map units. After further study, the geologist may decide that two or more nearby bodies of plutonic rock exposed on the Earth's surface are similar in all essential respects, including known or inferred age. Even though they may not be connected at the Earth's surface, the geologist may thus combine several masses of similar plutonic rock into a single geologic map unit, inferring that they are somehow connected below the surface and represent a single intrusive episode. This grouping of isolated bodies of related plutonic rock into a single geologic map unit is analogous to the grouping of discontinuous exposures of similar sedimentary rock into formations, such as the Coconino Sandstone and the Kaibab Limestone, which are well exposed in the Grand Canyon region. For ease of reference, the plutonic-rock units likewise are generally named for an appropriate geographic feature, plus a compositional term: for example, the El Capitan Granite, the Half Dome Granodiorite, and the granodiorite of Kuna Crest.

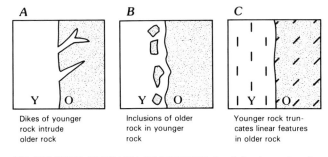

Dikes of younger rock intrude older rock

Inclusions of older rock in younger rock

Younger rock truncates linear features in older rock

FEATURES SEEN IN OUTCROP that help determine the relative ages of plutonic rocks: O, older pluton; Y, younger pluton. (Fig. 17)

GRANITIC ROCKS OF YOSEMITE

The plutonic rocks of Yosemite have been mapped and studied in considerable detail. Few of those details can be shown on the generalized geologic map in this volume (pl. 1), but a geologic map at a much larger scale is available (Huber and others, in press). On that map, the granitic rocks of the Yosemite area are separated into nearly 50 different plutonic-rock units, each consisting of one or more individual bodies of rock. An even larger scale geologic map is available for Yosemite Valley (Calkins, 1985; see section above entitled "Geologic Maps of Yosemite").

Some plutonic-rock units are further grouped into intrusive suites. The concept underlying an intrusive suite is that all the rocks in the suite resulted from the same magma-producing event. Geologists are most sure of a common ancestry if the rocks in a suite grade into each other. Such suites commonly are zoned, both compositionally and texturally, and generally exhibit partial or complete nested patterns in which relatively dark rock in the margins gives way inward to younger, lighter colored rock in the interior. The units that compose this ideal kind of intrusive suite are believed to result from modifications of a common parent magma. Examples include the Tuolumne Intrusive Suite, the first intrusive suite to be identified in the Sierra Nevada, and the intrusive suite of Buena Vista Crest. The geologic map (pl. 1) groups most plutonic-rock units into intrusive suites and thus provides a broader picture of the major pulses of plutonic activity that contributed to the construction of the Sierra Nevada batholith. The more detailed geologic map of Yosemite Valley (pl. 2) delineates not only intrusive suites but also component units of the suites.

All the plutonic rocks within Yosemite National Park proper are believed to be of Cretaceous age, with the possible exception of some small bodies of diorite and gabbro that may be somewhat older. Some Jurassic plutonic rock does occur just west of the park, west of the Big Oak Flat entrance, and some Triassic plutonic rock occurs east of the park in Lee Vining and Lundy Canyons. These rocks are included with "Plutonic rocks, unassigned to suites" on plate 1 and are shown individually only on the larger scale geologic map published separately (Huber and others, in press).

Examples of many of the named rock types in Yosemite are displayed at the Valley Visitor Center, where they may easily be compared; they are next described for two readily accessible areas in the park, Yosemite Valley and the Tuolumne Meadows area.

BIRD'S-EYE VIEW OF YOSEMITE VALLEY, with selected landforms identified. (Fig. 18)

YOSEMITE VALLEY AREA

The oldest plutonic rocks of the Yosemite Valley area compose the walls of Merced Gorge and the west end of the valley. They include the diorite of the Rockslides, the granodiorite of Arch Rock, and the tonalite of the Gateway (pl. 2). The largest outcrop of diorite is just west of the Rockslides (fig. 18), but the talus slopes below, composed of broken blocks of diorite, are more accessible. A good exposure of the granodiorite of Arch Rock can be seen immediately east of the Arch Rock Entrance Station on the El Portal Road (Route 140), where the road passes under two large fallen blocks of the granodiorite (park vehicles near the entrance station). The tonalite of the Gateway can be seen along the El Portal Road across from the first turnout after the road starts climbing up the Merced Gorge eastward from El Portal; these last two locations are west of the map area shown in plate 2. Studies of radiometric decay indicate that the tonalite of the Gateway is about 114 million years old. The radiometric age of the granodiorite of Arch Rock has not been determined, but it probably is only a little younger than that of the Gateway.

The El Capitan Granite subsequently intruded these older plutonic rocks about 108 million years ago and now makes up the bulk of the west half of the valley area. About 4 km east of the Arch Rock Entrance Station, the El Portal Road cuts through blocks of El Capitan Granite dislodged in a 1982 rockfall. These blocks, some the size of a small house, display fresh surfaces of the granite (fig. 10; see fig. 49), as well as numerous inclusions of dark-colored rock. The imposing monoliths of Turtleback Dome, El Capitan, Three Brothers, and Cathedral Rocks also are hewn chiefly from massive El Capitan Granite.

After the El Capitan Granite was emplaced, the Taft Granite welled up and intruded the El Capitan. Dikes of Taft Granite invading El Capitan Granite and inclusions of El Capitan in Taft establish the Taft as younger. The two rocks are similar, but Taft Granite is lighter in color and commonly finer grained than El Capitan Granite and, unlike El Capitan Granite, generally does not contain phenocrysts. Taft Granite forms the brow of El Capitan and part of the upland between El Capitan and Fireplace Bluffs. On the south side of the valley, Taft Granite can be seen at Dewey Point and near The Fissures, just east of Taft Point.

In the vicinity of Leaning Tower and Cathedral Rocks, dikes and irregular masses of several fine-grained rocks cut the Taft and El Capitan Granites. Examples of these fine-grained rocks can be seen in blocky rubble near the base of Bridalveil Fall. The Leaning Tower Granodiorite characteristically contains rounded clots of dark minerals that give it a spotted appearance (fig. 11C). The Bridalveil Granodiorite, which contains fine, evenly distributed, light and dark minerals, has a salt-and-pepper appearance (fig. 11D); features seen in outcrop show that it intruded nearly all the rocks which it now contacts.

Dark, fine-grained diorite also intrudes the El Capitan and Taft Granites. A striking example is exposed on the east face of El Capitan, where dikes of diorite form an irregular pattern that, in part, very crudely resembles a map of North America (fig. 19).

The Sentinel Granodiorite forms a north-south band that crosses the valley between Taft Point and Glacier Point. The rock varies in appearance but is generally medium gray and medium grained (fig. 11B). Giant inclusions of El Capitan Granite are embedded within Sentinel Granodiorite in a zone that extends along Yosemite Creek and down the face of the cliff near Yosemite Falls. The Sentinel Granodiorite reappears on the south valley wall west of Union Point and then extends southward through Sentinel Dome to Illilouette Ridge. Dikes of Sentinel Granodiorite

DIORITE DIKES on the face of El Capitan; dark patch is thought by some to resemble a crude map of North America. Some lighter colored dikes are also present. (Fig. 19)

that cut inclusions of El Capitan Granite can be seen in the roadcut along the Glacier Point Road near the trailhead to Taft Point.

The rock at Glacier and Washburn Points is darker than Sentinel Granodiorite and has a streaky appearance from parallel-oriented flakes of biotite and rods of hornblende. This darker rock, once thought to be part of the Sentinel and shown as such on earlier geologic maps, is now assigned to the granodiorite of Kuna Crest.

The Half Dome Granodiorite dominates the valley area east of Royal Arches and Glacier Point. It is medium to coarse grained and contains well-formed plates of biotite and rods of hornblende (fig. 11A). At Church Bowl and in the cliff west of Royal Arches, horizontal dikes of Half Dome Granodiorite cut the older granodiorite of Kuna Crest. Half Dome Granodiorite forms the sheer cliffs to the north of the trail between the Ahwahnee Hotel and Mirror Lake. The trail to Vernal and Nevada Falls also crosses through Half Dome Granodiorite. Except for minor dikes, the Half Dome Granodiorite, about 87 million years old, is the youngest plutonic rock in the valley area.

TUOLUMNE MEADOWS AREA

The granodiorite of Kuna Crest and the Half Dome Granodiorite exposed at the east end of Yosemite Valley are two plutonic-rock units that make up the western margin of the Tuolumne Intrusive Suite. This suite underlies a large part of eastern Yosemite National Park from upper Yosemite Valley, across Tuolumne Meadows eastward to the crest of the Sierra, and northward beyond the park boundary (pl. 1). The Tuolumne Intrusive Suite, one of the best studied groups of granitic rocks in the Sierra, consists of four bodies of plutonic rock, sequentially emplaced and partly nested one within the other (fig. 20). The suite is well exposed in the area centered on Tuolumne Meadows, and the Tioga Road (Route 120) provides access to many conspicuous outcrops of the suite's components.

The oldest and darkest plutonic rock generally forms the margin of the suite, and the youngest rock is in its core. The rocks are, from oldest to youngest: the granodiorite of Kuna Crest (about 91 million years old), the Half Dome Granodiorite, the Cathedral Peak Granodiorite (about 86 million years old), and the Johnson Granite Porphyry. Field relations indicate that the Johnson Granite Porphyry is the youngest granitic rock in the park, although a radiometric age has not

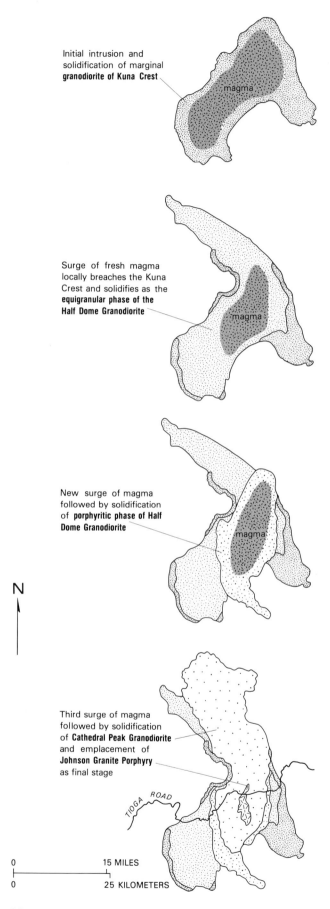

Initial intrusion and
solidification of marginal
granodiorite of Kuna Crest

Surge of fresh magma
locally breaches the Kuna
Crest and solidifies as the
**equigranular phase of the
Half Dome Granodiorite**

New surge of magma
followed by solidification
of **porphyritic phase of Half
Dome Granodiorite**

Third surge of magma
followed by solidification
of **Cathedral Peak Granodiorite**
and emplacement of
Johnson Granite Porphyry
as final stage

N

0   15 MILES

0   25 KILOMETERS

yet been determined. The granodiorite of Kuna Crest normally occupies the margin of the suite, but on much of the perimeter the Half Dome Granodiorite and the Cathedral Peak Granodiorite have broken through the granodiorite of Kuna Crest to form the marginal units (fig. 20).

The overall concentric zonation of rock bodies within the suite, as well as the overall chemical similarities among the rocks, suggests that these rocks originated from the same magma chamber. This inferred common parentage provides the rationale for grouping these rocks into an intrusive suite. The composition of the magma, however, changed over time: the older, hornblende- and biotite-rich rocks at the margins give way to quartz- and potassium feldspar-rich rocks toward the center. Hornblende and biotite crystallize at higher temperatures than quartz and feldspar, and so during cooling of a magma, these dark minerals generally crystallize earlier than the light-colored ones. This relation suggests that cooling of the magma started at the margins and progressed inward over time.

North of the Tioga Pass Entrance Station, the trail to Gaylor Lakes crosses over the granodiorite of Kuna Crest, the oldest and darkest rock in the Tuolumne Intrusive Suite. This trail weaves back and forth near the contact between the granodiorite and the metamorphic rocks that it intruded. The granodiorite also contains many disc-shaped inclusions that are oriented parallel to its contact with the older metamorphic rocks. These inclusions were probably stretched and oriented by movement within the magma during intrusion and cooling.

The Half Dome Granodiorite, the next youngest rock in the suite, is in contact with the granodiorite of Kuna Crest to the west along the ridge crossed by the Gaylor Lakes Trail. The best exposures of the Half Dome, however, are surrounding the turnout at Olmsted Point west of Tenaya Lake. Fresh, clean outcrops of the rock abound at and across from the turnout. Half Dome Granodiorite makes up much of the southwestern part of the Tuolumne Intrusive Suite and in several areas is the marginal rock.

Heading east toward Tuolumne Meadows, the Tioga Road crosses the contact between the Half Dome Granodiorite and the Cathedral Peak Granodiorite just east of Tenaya Lake. The contact is obscure, however, because here the Half Dome contains nearly as many potassium feldspar phenocrysts as does the younger

EVOLUTION OF THE TUOLUMNE INTRUSIVE SUITE—a map view. (Fig. 20)

Cathedral Peak. Pothole and Lembert Domes, both marginal to the meadows, are composed entirely of Cathedral Peak Granodiorite. The rock of these domes clearly displays potassium feldspar phenocrysts, commonly as much as 2 to 3 in. long (fig. 14). These impressive crystals stand out against a medium-grained background. The Cathedral Peak Granodiorite forms the largest pluton of the Tuolumne Intrusive Suite, extending long distances to the north and south of Tuolumne Meadows.

The youngest, smallest, and most central rock body of the suite is composed of the Johnson Granite Porphyry. In a *porphyry*, the conspicuous phenocrysts are set in a finer grained matrix than in such porphyritic rocks as the Cathedral Peak Granodiorite, and so individual mineral grains in the matrix are difficult to identify without a microscope. Low outcrops of the porphyry can be seen in Tuolumne Meadows along the Tuolumne River, across from the store, and east of Soda Springs on the north side of the river. The rock is very light colored, with only a few scattered potassium feldspar phenocrysts within a fine-grained matrix (fig. 21). Dikes of Johnson Granite Porphyry intrude Cathedral Peak Granodiorite, and the porphyry itself is cut by light, fine-grained aplite dikes.

The fine-grained matrix of a porphyry requires that

JOHNSON GRANITE PORPHYRY, showing potassium feldspar phenocrysts set in a fine-grained matrix. (Fig. 21)

partially crystallized magma be quenched or cool relatively quickly. Such conditions would result from a sudden release of pressure, as would occur if some of the magma were erupted at the Earth's surface. Thus, volcanic eruptions probably accompanied final emplacement of the Tuolumne Intrusive Suite—a volcanic caldera may once have existed far above what is now Johnson Peak (fig. 22).

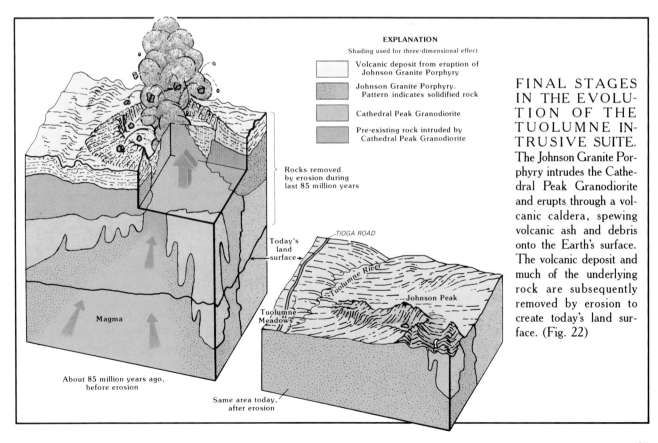

EXPLANATION

Shading used for three-dimensional effect

Volcanic deposit from eruption of Johnson Granite Porphyry

Johnson Granite Porphyry. Pattern indicates solidified rock

Cathedral Peak Granodiorite

Pre-existing rock intruded by Cathedral Peak Granodiorite

Rocks removed by erosion during last 85 million years

Today's land surface

TIOGA ROAD

Tuolumne River

Johnson Peak

Tuolumne Meadows

Magma

About 85 million years ago, before erosion

Same area today, after erosion

FINAL STAGES IN THE EVOLUTION OF THE TUOLUMNE INTRUSIVE SUITE. The Johnson Granite Porphyry intrudes the Cathedral Peak Granodiorite and erupts through a volcanic caldera, spewing volcanic ash and debris onto the Earth's surface. The volcanic deposit and much of the underlying rock are subsequently removed by erosion to create today's land surface. (Fig. 22)

21

Metamorphic Rocks—Ancient Sediment and Lavas

Metamorphic rocks are derived from preexisting rocks by mineralogic and structural changes in response to increases in temperature, pressure, and shearing stress at depth within the Earth's crust. In the Sierra Nevada, some of this heat and pressure was supplied by the intruding granitic rocks, but much of it was imposed simply by depressing sedimentary and volcanic rocks once exposed at the Earth's surface downward to depths where higher temperature and pressure are the normal environment. The metamorphic rocks in the Yosemite area were derived from a great variety of sedimentary and volcanic rocks and thus exhibit a great variety in themselves. Some rocks have been only mildly metamorphosed and still retain original structures, such as sedimentary layering, that help to identify the nature of the original rock. Others have been so strongly deformed and recrystallized that original textures and structures have been destroyed, and determination of the original rock type is difficult.

Metamorphosed sedimentary rocks in the Yosemite area include rocks that were originally sandstone and siltstone, conglomerate, limestone, shale, and chert. Metamorphosed volcanic rocks in the Yosemite area include those derived from lava flows and various types of pyroclastic rocks—those formed from violently erupted volcanic debris.

The rocks into which the Sierra Nevada batholith was emplaced are weakly to strongly metamorphosed, mildly to complexly deformed strata of probable Paleozoic and Mesozoic age. In the Yosemite area these metamorphic rocks occur in two northwest-trending belts situated largely east and west of the park proper and in small isolated bodies scattered throughout the park. Fossils are scarce, and the radiometric ages of most of these rocks are poorly known.

Rocks of the western metamorphic belt underlie much of the foothills of the western Sierra between the San Joaquin and Feather Rivers, and form the western wallrocks of the Sierra Nevada batholith. In the canyon of the Merced River approaching Yosemite on Route 140, strikingly banded chert is exposed in the vicinity of the "geological exhibit" and eastward for several miles (fig. 23). This banded chert was formed from the skeletons of very tiny, silica-secreting marine animals called radiolarians; upon the death of such animals, their skeletons settle to the ocean bottom, where they collect in enormous numbers. Although the chert beds are moderately to strongly deformed, the

CONTORTED CHERT BEDS along the Merced River west of El Portal are ancient marine sediment that has been metamorphosed. (Fig. 23)

rock is easily recognizable as of sedimentary origin. In contrast, metamorphic rocks just west of El Portal and just west of Crane Flat along the Big Oak Flat Road (Route 120) have a metamorphic layering that largely destroys original bedding, and the origin of these rocks as sediment is less obvious. Fossils in a limestone bed just west of the "geological exhibit" on Route 140 indicate a Triassic age for at least some of the rocks exposed along this part of the Merced River canyon.

The eastern belt of metamorphic rocks extends for about 50 mi from south of Mammoth Lakes to north of Twin Lakes (pl. 1). Furthermore, rather than bounding the batholith, this belt is a giant septum of metamorphic rocks separating plutonic rocks on either side.

This eastern belt includes rocks of both sedimentary and volcanic origin, which range in age from early Paleozoic to late Mesozoic. The Paleozoic rocks are metasedimentary and include such varieties as quartzite, metaconglomerate, and marble. The commonest rock, however, is hornfels—a catchall term for a fine-grained metamorphic rock composed of a mosaic of equidimensional grains formed by recrystallization of sedimentary and volcanic rocks of various compositions. These Paleozoic rocks are well exposed along Route 120 near Ellery Lake east of Tioga Pass.

The Mesozoic rocks of the eastern metamorphic belt are chiefly of volcanic origin—tuff and other explosively ejected fragmental volcanic rock—with lesser amounts of sedimentary rock. These Mesozoic rocks, which lie generally west of the Paleozoic rocks in the eastern metamorphic belt, make up the Ritter Range and the southeastern margin of the park, and much of the Sierran Crest northward through Kuna Peak, Mount Dana, Gaylor Peak, and continuing north of the park beyond Twin Lakes (pl. 1). Relict sedimentary bedding is commonly preserved—steeply dipping, as west of Saddlebag Lake, or highly contorted, as near Spotted Lakes at the south end of the park (fig. 24).

Of particular interest are the little-deformed metamorphic rocks of Cretaceous age. Metamorphosed volcanic rocks near the summit of Mount Dana have a radiometric age of about 118 million years, and those from the Ritter Range of about 100 million years, which means that their eruption from volcanoes occurred at the same time that some of the smaller plutonic-rock suites were emplaced at depth. In the Ritter Range, a thick deposit of volcanic breccia has been interpreted as resulting from collapse of an ancient volcanic caldera.

LATE CENOZOIC VOLCANIC ROCKS— BORN OF FIRE

Volcanic rocks, like their plutonic counterparts, are also classified on the basis of composition. Because volcanic rocks erupted onto the Earth's surface cool and solidify more quickly than plutonic rocks, they tend to be finer grained or even glassy, with few minerals identifiable to the eye. Those few visible minerals, however, are guides to the rock's composition. Late Cenozoic volcanic rocks in Yosemite have a very limited range in composition; they generally contain little or no quartz and range from basalt and andesite (containing little or no potassium feldspar) to latite (containing both potassium and plagioclase feldspar). A volcanic rock containing quartz—rhyolite—does occur just east of Yosemite at the Mono Craters.

Late Cenozoic volcanic rocks of the Yosemite area formed both by the eruption of vast volumes of lava and by much smaller eruptions. The products of great eruptions extend into the northern part of the park but are much more extensive in the northern Sierra; they include lava flows, tuff, and volcanic mudflows. Details of the nature and distribution of all these volcanic rocks are deferred to the section dealing with the late Cenozoic.

A

B

METAMORPHIC ROCKS with relict sedimentary bedding. *A*, Steeply dipping, as northwest of Saddlebag Lake. *B*, Highly contorted, as near Spotted Lakes. Photograph by John P. Lockwood. (Fig. 24)

The Earth is generally depicted as consisting of a series of concentric shells—a relatively thin outer *crust*, an intermediate *mantle*, and an interior *core* (fig. 25). The Earth's crust and uppermost part of the upper mantle together form the rigid outer part of the Earth—the *lithosphere*—which is broken into plates that ride over a less rigid, viscous layer within the upper mantle that yields plastically. There are seven very large plates, and a dozen or so small ones (not all of which are shown in fig. 26.) The large plates consist of both oceanic and continental portions; the present North American plate, for example, includes not only the North American Continent but Greenland and the west half of the North Atlantic Ocean as well. The crust beneath continents is typically 20 to 34 mi thick and is less dense than the crust beneath oceans, which typically is only 4 to 5 mi thick. The plates generally are internally rigid, and most dynamic geologic activity is concentrated along the plate boundaries; these boundaries are marked by long, narrow belts of earthquake and volcanic activity.

Each of the plates is moving relative to all the others. In the simplest mode, two plates slide past each other along a strike-slip fault (fig. 27*A*). The San Andreas fault, running much the length of California and forming part of the present boundary between the North American and Pacific plates, is an example. Where plates move away from each other, primarily along the system of great submarine ridges in the world's oceans, hot material wells up from below to fill the gap (fig. 27*B*). As this hot material cools to form basalt, it becomes attached to the plates on either side of the spreading zone, and new crust is created. Where plates converge, one tips downward and slides beneath the other— a process called *subduction* (fig. 27*C*). Generally, a plate with dense oceanic crust slides beneath one with more bouyant continental crust. Thus, new oceanic crust created at spreading centers is recycled back into the Earth's interior through subduction, and and the total surface area of the Earth remains unchanged.

PLATE TECTONICS— A DYNAMIC GLOBE

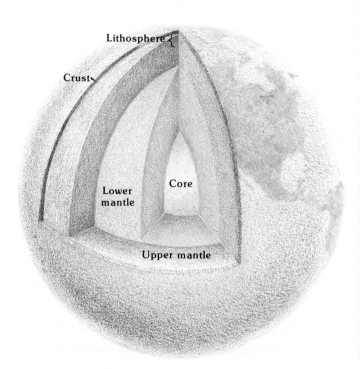

INTERIOR OF THE EARTH, showing relation of crust and mantle to the rigid lithosphere—the stuff of which the mobile plates are made. (Fig. 25)

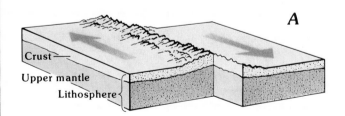

A

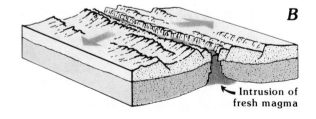

B

MAJOR LITHOSPHERE PLATES OF THE
WORLD, showing boundaries that are presently active. Double
line, zone of spreading, from which plates are moving apart;
barbed line, zone of underthrusting (subduction), where one plate
is sliding beneath another—barbs on overriding plate; single line,
strike-slip fault, along which plates are sliding past one another.
See figure 27 for examples of plate motions. (Fig. 26)

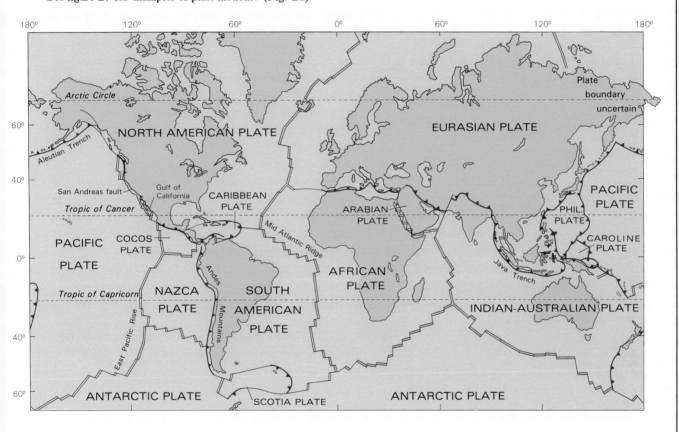

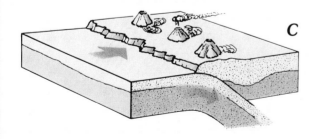

THREE PRINCIPAL KINDS OF PLATE MOTION.
A, The plates slide past each other along a strike-slip fault. *B*,
The plates move away from each other at a divergent boundary. *C*,
The plates move toward each other at a convergent boundary; the
process of subduction consumes crust at convergent plate bound-
aries. (Fig. 27)

GENESIS OF YOSEMITE'S ROCKS

The geologic story of Yosemite as presented up to this point has been largely a description of the rocks as we see them now. But how did they get this way? And when? The search for answers to these questions involves interpretation of geologic observations made in Yosemite and elsewhere in the Sierra Nevada, together with numerous inferences based on accumulated geologic knowledge and on theoretical concepts. Some parts of the geologic history can be deciphered with confidence and in considerable detail, but other parts are less complete because the geologic data are very spotty.

A SINGLE QUIET PLATE— THE PALEOZOIC

The framework within which most geologists today view geologic processes, such as the creation of batholiths and the building of mountains, is that of the theory of *plate tectonics* (see section above entitled "Plate Tectonics — ***"). Tectonics is the study of the deformation of earth materials and the structures resulting from that deformation. The "tectonics" in plate tectonics refers to deformation and structure on a global scale.

The oldest rocks in the Yosemite area were derived from sediment deposited during early Paleozoic time, beginning about 500 million years ago. During the Paleozoic, the area that was to become Yosemite was near the west edge of the growing North American Continent. The setting, for the most part, was a relatively passive one. Paleozoic sediment derived by erosion of still older rocks to the east was delivered by ancient streams flowing westward to a sea along the continental margin. Deposition of such sediment throughout most of the Paleozoic, though not necessarily continuous, resulted in the accumulation of thousands of feet of mud and sand, which eventually consolidated into shale and sandstone. Plant and animal life in the sea contributed their part by depositing calcium carbonate and silica, later to become beds of limestone and chert.

During the Paleozoic, the continent and its adjacent sea appear to have been traveling together on a single plate. All was not totally passive, however, because there is evidence for folding and deformation of some early Paleozoic strata during the late Paleozoic. It is not possible to relate such deformation to specific plate-margin tectonics because of severe overprinting by later tectonic events. By the end of the Paleozoic the geometry at the west edge of the North American plate had changed, and an oceanic plate was now underriding, or being subducted beneath, the North American plate.

A TIME OF FIRE AND UPHEAVAL— THE MESOZOIC

The presence of a subduction zone along the west margin of the North American plate had profound effects on that plate. As the cool oceanic plate was subducted, the overriding continental plate was deformed. But more important to the Yosemite story were the igneous effects of subduction. Wherever convergent plate margins and subduction zones are present today, magma is generated at depth, and linear belts of volcanoes form atop the overriding plate, parallel to the subduction zone. Mount St. Helens, for example, and other volcanoes of the Cascade Range lie parallel to an active subduction zone that extends from northern California to Canada, and we infer that ancient subduction zones produced similar belts of igneous activity.

We can only speculate as to the nature of the physical and chemical processes that take place within a subduction zone. A prevalent theory is based on experiments indicating that the presence of water lowers the melting temperature of rock materials. This theory holds that water entrapped in the descending slab of oceanic crust is driven out as the slab reaches higher temperatures and leaks upward into the overriding lithosphere, where partial melting results (fig. 28). Magma generated in the mantle part of the lithosphere has the composition of basalt or andesite, but as the magma rises into the continental crust, a more silicic magma may be generated—one with the composition of rhyolite or granite. After rising toward the Earth's surface, this silicic magma may erupt as rhyolite volcanoes, or cool and come to rest as great bodies of granitic rock within the upper crust. Most geologists now believe that this is the mechanism—greatly simplified here — through which the Sierra Nevada batholith was generated and emplaced.

By early Mesozoic time, more than 200 million years ago, magma reached the Earth's surface in a belt of volcanoes and spewed forth to form great volumes of volcanic rock, metamorphosed remnants of which are now exposed in the area of the Sierran crest (pl. 1). By this time, silicic magma had also formed, some of which cooled and solidified below the Earth's surface to form bodies of granitic rock; one such body is now exposed in Lee Vining Canyon (Scheelite Intrusive Suite, pl. 1). Subduction along the margin of the North American plate was not continuous during the Mesozoic, and subsequent movement of granitic magma into the upper crust was somewhat episodic; the greatest volumes were emplaced during the middle Jurassic and Late Cretaceous. By the beginning of the Cenozoic, the magmatic system in the Sierran region shut off, leaving behind the mass of granitic rock we now call the Sierra Nevada batholith.

Emplacement of plutonic rocks within the upper crust was probably accompanied by many contemporaneous volcanic eruptions at the Earth's surface. Evidence in the Yosemite area for such eruptions includes the texture of the Johnson Granite Porphyry (fig. 21) and similar porphyries in other intrusive suites, and the 100- to 118-million-year ages of the volcanic rocks near Mount Dana and in the Ritter Range. In addition, volcanic eruptions associated with emplacement of the Sierra Nevada batholith and other contemporaneous batholithic complexes now exposed along the western margin of the North American Continent provide the only apparent source for the extremely voluminous deposits of Cretaceous volcanic ash to the east in the continental interior.

Not all of the oceanic plate was being subducted during that time, however. Parts of that plate, particularly the upper layer of marine sedimentary rocks on the oceanic crust, were added, or accreted, to the leading margin of the overriding continental crust. The banded chert in the Merced River canyon west of El Portal, once part of an ocean floor, was added to the North American plate by such a process.

The end result of the intrusion of the batholith, the construction of volcanoes, and the deformation of the metamorphic rocks was a linear mountain range parallel to and inboard of the continental margin. This range has been referred to as the ancestral Sierra Nevada. Mountains are born, only to be worn down by erosion; and erosive forces begin to act even as the mountains are being upraised. Nevertheless, the ancestral Sierra probably reached elevations above 13,000 ft, similar to those in the Cascade Range in western Washington and Oregon, a range being constructed over an active subduction zone today.

What caused magmatism in the Sierra to cease during the late Mesozoic? Many geologists speculate that the subducting oceanic slab speeded up and flattened out, so that the zone of magma generation shifted eastward. Although there are no giant batholiths in Nevada, many bodies of granitic and volcanic rock occur there that are chiefly of Cenozoic age, younger than the Sierra Nevada batholith.

Once the magmatic construction of the ancestral Sierra Nevada ceased, erosion became the dominant force in shaping the range, mostly by removing it. Before the end of the Mesozoic, some 63 million years ago, the volcanoes had largely been removed, and the batholith itself was exposed and being eroded. Sediment derived from this erosion was transported by streams coursing down the slope of the range to the Central Valley, where it now forms deposits as much as tens of thousands of feet thick. By middle Cenozoic time, so much of the range had been removed that it had a relief of only a few thousand feet or so.

THE SIERRA GROWS AGAIN— THE LATE CENOZOIC

During early Cenozoic time the Sierra Nevada region was relatively stable, and the range continued to be worn down faster than it was rising. But during the late Cenozoic, from about 25 to 15 million years ago, a dramatic change in plate motion along the edge of the North American plate occurred, with far-reaching effects. The oceanic plate that was being subducted beneath the Sierra Nevada was totally consumed into

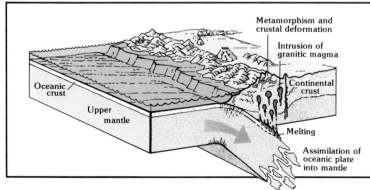

SUBDUCTION OF AN OCEANIC PLATE during convergence with a continental plate. Magma, formed by partial melting of overriding continental plate, rises into continental plate to form volcanoes and plutons along a mountain chain. (Fig. 28)

the subduction zone, and the plate that replaced it was moving in a different direction—northwesterly. The boundary between the North American plate and this northwesterly-moving plate, called the Pacific plate, became a strike-slip fault along this segment of California—the San Andreas fault (fig. 26).

This change in plate-boundary motion, from convergence to lateral motion, caused a change in the pattern of stresses imposed on the Sierran region. The continental crust east of the Sierra began to expand in an east-west direction, and the thick, light-weight Sierran crust began to rise again. The exact mechanism of this uplift is not understood, but the results are there to see. In the Yosemite area, the Sierra is clearly an uptilted block of the Earth's crust, with a long slope westward to the Central Valley and a steep escarpment separating it from the country to the east (fig. 29). Total uplift in the vicinity of Mount Dana during late Cenozoic time to the present is estimated at about 11,000 ft.

The uplift began slowly and accelerated over time. The range certainly is still rising—and the rate may still be accelerating. The estimated current rate of uplift at Mount Dana, less than 1½ inches per 100 years, may appear small, but it is greater than the overall rate of smoothing off and lowering of the range by erosion. Thus, there is a net increase in elevation. Estimates of uplift amount and rate are based on studies of lava flows and stream deposits thought to be nearly horizontal when formed, but which are now tilted westward toward the Central Valley. Progressive tilt is indicated by older deposits with greater inclinations than younger ones.

François Matthes inferred from his studies that the late Cenozoic uplift occurred in a series of three pulses, interrupted by pauses in uplift. In his view, each pulse initiated a new cycle of erosion and thus produced a stage of landscape incision characterized by successively greater relief: Matthes' broad-valley, mountain-valley, and canyon stages. More recent studies show that

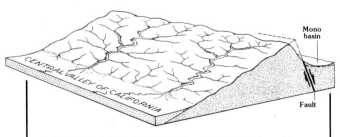

UPLIFT AND TILT of the Sierran block, with east escarpment formed along fault. Arrows show direction of movement on fault. (Fig. 29)

fortuitous correlation and the commonly local control of erosion weaken Matthes' case for three distinct pulses of uplift. This does not mean that the uplift was entirely uniform—few things in geology are—but rather that uplift, once initiated, was more nearly continuous than he envisioned.

At the same time that the Sierra was undergoing uplift and erosion and incision by streams, volcanoes again became active in parts of the range, particularly north of Yosemite. During the interval from about 20 million years ago to about 5 million years ago, vast volumes of volcanic material were erupted from a belt of volcanoes extending along what is now the Sierra crest north of Yosemite. These volcanoes were the southward extension of the Cascade Range of volcanoes still active in northern California, Oregon, and Washington. With the advent of the San Andreas strike-slip fault, the subduction complex associated with Cascade volcanism migrated northward, and the Sierran volcanoes turned off. Lassen Peak in northern California is the southernmost volcano of this chain that is still active.

During this late Cenozoic volcanism, the Sierra Nevada north of Yosemite was virtually buried by lava flows, volcanic tuff, and volcanic mudflows. The volcanic material traveled great distances. Much of it reached the margin of the Central Valley, and some of it traveled as far south as the northern part of Yosemite. Three separate units of this volcanic extravaganza—a mudflow, a lava flow, and a volcanic tuff—successively flowed down the valley of a south-flowing tributary of the ancestral Tuolumne River and into the main channel in the vicinity of Rancheria Mountain northeast of Hetch Hetchy (figs. 30, 31). Erosional remnants of the volcanic mudflow indicate that it flowed almost as far west as Groveland, some 20 mi west of the park. Other erosional remnants of this mudflow indicate that it was so thick that it actually flowed upstream along the ancestral Tuolumne River at least 5 mi above the junction of the south-flowing tributary.

Other volcanic rocks in Yosemite represent local eruptive events. One such event is recorded by a basalt plug—a solidified remnant of lava in a volcanic conduit—locally known as "Little Devils Postpile" and located on the south side of the Tuolumne River several miles west of Tuolumne Meadows. The outcrop, easily reached by the Glen Aulin trail, exhibits crudely developed columnar joints (fig. 32). This basalt is about 9 million years old.

One small lava flow of basalt, about 3½ million years old, was erupted just south of Merced Pass, and a few

ANCIENT CHANNEL OF THE TUOLUMNE RIVER on Rancheria Mountain northeast of Hetch Hetchy. The river flowed westward away from the viewer into the V-shaped notch cut into granite in the center of the photograph. About 10 million years ago, the channel and about 50 ft of river gravel were buried beneath a volcanic mudflow, the material seen on the slope above the "V". Its cross section now exposed by erosion, this ancient channel was first described as such by Henry W. Turner, who took this photograph about 1900. (Fig. 30)

VOLCANIC MUDFLOW DEPOSIT on Rancheria Mountain. Brown, smooth-appearing slope to the right, at middle distance, is underlain by a volcanic mudflow deposit filling an ancient stream channel. This photograph, which is another view of the channel in figure 30, shows its position on the slope above Piute Creek, which drains away from the viewer to the present canyon of the Tuolumne River in the background. The former Tuolumne River flowed from left to right into the base of the brown area, concealed by trees in the center of the photograph, and at this point was more than 1,500 ft above the present canyon of the Tuolumne. This difference in elevation indicates the amount of stream incision by the Tuolumne River since its former channel was filled and abandoned and the river was forced to cut a new one. Photograph by Clyde Wahrhaftig. (Fig. 31)

scattered flows of similar age lie just south and southeast of the park. These flows record the most recent igneous activity in Yosemite.

Things have not remained quiet east of Yosemite, however. A cataclysmic eruption about 700,000 years ago created 10- by 20-mi-wide Long Valley caldera, within which now sits the town of Mammoth Lakes. This eruption spewed forth 2,500 times as much ash as the 1980 Mount St. Helens eruption; layers of the ash from Long Valley caldera have been found as far east as Nebraska. Volcanic rocks of Mammoth Mountain and the basalt at the Devils Postpile were erupted subsequently. The Mono Craters and Inyo domes between Mono Lake and Mammoth Lakes have been erupting episodically during the past few thousand years, and the most recent domes were formed only about 600 years ago. Such activity is almost certainly not yet finished.

COLUMNAR JOINTS in a basalt plug—a remnant of a volcanic conduit—at "Little Devils Postpile," adjacent to the Tuolumne River west of Tuolumne Meadows. (Fig. 32)

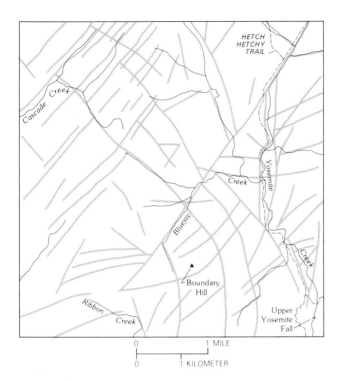

FINAL EVOLUTION OF THE LANDSCAPE

THE ROLE OF JOINTS

The bedrock structures having the greatest effects on Yosemite's landform development are *joints*. Joints are more or less planar cracks commonly found as sets of parallel fractures in the rock. Regional-scale joints commonly determine the orientation of major features of the landscape, whereas outcrop-scale joints determine the ease with which rock erodes. Joints are of overwhelming influence on landform development in granitic terrane because they form greatly contrasting zones of weakness in otherwise homogeneous, erosion-resistant rock and are avenues of

REGIONAL JOINTS emphasized by vegetation, as seen on an aerial photograph—Yosemite Creek basin north of Yosemite Valley. Northeast-trending joint set is more closely spaced than northwest-trending set. Sparser, east-west-trending set is also present. Accompanying sketch map shows orientation of major joints and illustrates the significant control of stream courses by joints. (Fig. 33)

access of water and air for weathering. In metamorphic rocks, such planar structures as bedding or aligned minerals commonly determine the orientation of fractures.

A regional system of widely spaced master joints is conspicuous throughout the granitic terrane, particularly in the High Sierra where rock exposures are extensive. These joints generally are nearly vertical and are not to be confused with the gently dipping sheet joints that are subparallel to topographic surfaces. Linear depressions commonly follow the master joints, and even in highly dissected regions, straight segments of streams are joint controlled. Generally, two principal sets of joints can be identified nearly at right angles, one set commonly trending northeast and the other northwest (fig. 33). The orientation changes from place to place, but most joints are straight or only gently curved. Some individual joints can be traced for many miles. The continuity of joint sets across the boundaries between individual granitic plutons indicates that the joints formed after consolidation of the entire batholith and thus are not cooling fractures of the individual bodies. The regional joint system evidently resulted from stresses imposed on the batholith during later tectonic events, such as tilting of the Sierra region.

Some movement other than simple opening along the joint planes has probably taken place along most of the master joints. Such lateral, or fault, movement, even though slight, would crush or break rock along the joint. Deep weathering and erosion along these zones of broken rock form long, linear depressions, many of them now channels for streams. Areas of granite within blocks bounded by master joints are themselves jointed to a lesser degree but remain more

STAIRCASE FALLS. Stairtreads follow joints inclined eastward. Photograph by Tau Rho Alpha. (Fig. 35)

cohesive and form the bolder areas between master joints.

On a more local scale, vertical joint sets are responsible for the orientation of major features, such as the planar face of Half Dome and the series of parallel cliffs at Cathedral Rocks. An individual cliff face itself may not be the original controlling joint surface, because material is continually spalling off, but its orientation is controlled by a preexisting joint.

Although vertical master joints prevail, inclined joint sets have added to the diversity of Yosemite's landforms. The west faces of the Three Brothers in Yosemite Valley were largely determined by a set of master joints that slope about 45° westward (fig. 34). The slope between Cathedral Rocks and Bridalveil Creek also follows a set of westward-inclined joints. The stairtreads of Staircase Falls follow east-dipping joints (fig. 35).

At outcrop scale, two nearly vertical joint sets, perpendicular to each other, combine with nearly hori-

INCLINED JOINTS determine westward slope of upper surfaces of the Three Brothers. Photograph from National Park Service collection. (Fig. 34)

31

RECTANGULAR BLOCKS formed in El Capitan Granite by intersecting joints. (Fig. 36)

zontal joints to create approximately rectangular blocks (fig. 36). Generally, the more siliceous, or quartz-rich, rocks (granite and granodiorite) have more widely spaced joints than the less siliceous rocks (tonalite and diorite). Also, the finer grained rocks generally have more closely spaced joints than the coarser grained ones. Thus, both composition and texture influence the spacing of joints in a given rock mass.

The least siliceous of the plutonic rocks in Yosemite Valley, for example, is the diorite that occurs at the Rockslides; this rock is also generally finer grained than most of the granitic rocks. The diorite is the most closely jointed rock in the valley, and enormous piles of joint-derived blocks of diorite have accumulated throughout the area of the Rockslides (fig. 37).

In a particularly striking contrast to the jumbled piles of debris at the Rockslides is the largely unbroken face of El Capitan, a short distance to the east—one of the sheerest cliffs in the world. It consists chiefly of El Capitan and Taft Granites, two of the most siliceous plutonic rocks in this area. The composition of these rocks determines the characteristics of El Capitan itself—massiveness and resistance. Because El Capitan is largely unjointed, the talus pile at its foot is small. Cathedral Rocks and the Leaning Tower are composed of El Capitan Granite, complexly intruded by Bridalveil

THE ROCKSLIDES AND EL CAPITAN. The Rockslides (left) is a jumbled collection of talus blocks of diorite, the most closely jointed rock in Yosemite Valley. In contrast, El Capitan (right) is largely unjointed granite, and the pile of debris at its foot, though concealed in this photograph, is comparatively small. (Fig. 37)

Granodiorite; these pinnacles stand out as they do because of their siliceous composition.

The apron or lower part of the cliff east of Glacier Point consists of unjointed Half Dome Granodiorite capped by well-jointed tonalite; from a viewpoint in Stoneman Meadow, an observer, by noting this difference in structure, can trace the contact between the two rock types quite closely (fig. 38).

So mammoth a feature as Half Dome could only have been carved from a sparsely jointed rock. In spite of James Hutchings' eloquent statement accompanying the frontispiece to this volume, Half Dome is nearly as whole as it ever was. The impression from the valley floor that this is a round dome which has lost its northwest half is an illusion. From Glacier Point or, even better, from Washburn Point, we can see that it is actually a thin ridge of rock oriented northeast-southwest, with its southeast side almost as steep as its northwest side except for the very top. Although the trend of this ridge, as well as that of Tenaya Canyon, is probably controlled by master joints, 80 percent of the northwest "half" of the original dome may well still be there. What probably happened is that frost splitting of the rock at the back of a tiny glacier against Half Dome above Mirror Lake gradually quarried back the steep northwest face. As the base of the cliff was hewn away,

ultimately parts of the sheets parallel to the original upper surface of Half Dome were left projecting outward at the crest of the vertical cliff. Sharp angular bends in the gross form of Yosemite Valley suggest that the entire valley, as well as Tenaya Valley, may have been eroded along a complex joint system now concealed by stream deposits on the valley floor.

The type of jointing that has most influenced the form of Yosemite's landmarks, however, is the broad, shell-like *unloading joints* or *sheeting,* also commonly referred to as *exfoliation.* Granitic rocks crystallize at considerable depth within the Earth while under great pressure from miles of overlying rock. As the still-buried plutonic rocks are uplifted into mountains and the overlying rock is eroded, the unloading, or release of the previously confining pressure, causes the rock to expand toward the Earth's surface. In jointed rocks, such expansion is taken up by adjustments along the numerous partings; but in a massive monolith, the stresses accumulate until they exceed the tensile strength of the rock, and the outer and more rapidly expanding layer bursts loose. Over time, the process is repeated, and the monolith becomes covered with several layers of shells. The outermost layer, exposed to the weather, gradually disintegrates, and the pieces fall off. The process of

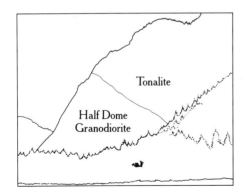

BOLD EXPOSURE of unjointed Half Dome Granodiorite (in sun) capped by mostly well-jointed tonalite (in shade) making up Glacier Point. The contact between the two rock types angles upward to left. (Fig. 38)

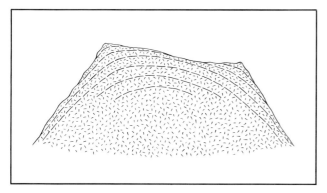

PROGRESSIVE ROUNDING OF MASSIVE GRAN-
ITE as successive sheets (dashed lines) are formed and spalled off
in response to unloading, or release of the confining pressure under
which the granite crystallized deep within the Earth. This is a form
of exfoliation. (Fig. 39)

A

B

C

D

sheeting eliminates projecting corners and angles and
replaces them with curves (fig. 39). As succeeding
shells drop off, these curves become more and more
gentle, and thus a smoothly rounded surface evolves.

Because the expansion that forms sheet joints
takes place perpendicular to the Earth's surface, the
shape of sheets generally reflects the topography,
although their formation subtly modifies the topography at the same time. If the ground surface is level,
the sheets will be horizontal (fig. 40A). If the granite
underlies a hill, the sheets will curve accordingly,
convex upward (fig. 40B); and if beneath a valley,
concave upward (fig. 40C). Sheeting also tends to
parallel the walls of canyons. If the canyon wall
slopes toward the river, the sheets do also. If the
walls are vertical, the sheets are also vertical (fig.
40D); thus, the vertical cliffs of Yosemite that appear
to be unbroken monoliths may have hidden vertical
fractures behind and parallel to the cliff face. The
undulating surface of the wall below Clouds Rest is
an outstanding example of sheeting that parallels the
topographic surface; the sheets are concave in the
bowl-shaped basins high on the cliff face and convex
on the intervening spurs (fig. 41).

The Royal Arches is a gigantic expression of sheet
jointing, with sheets as much as 200 ft thick (fig.
42). Too far below the surface to form the tops of
domes, the arches reveal a cross-sectional view of
sheet jointing in Half Dome Granodiorite that has
been truncated by the north wall of Yosemite Valley.
Similar features can also be seen on the walls behind
Upper Yosemite Fall and at the head of Ribbon Fall
alcove.

SHEET JOINTS FOLLOW TOPOGRAPHIC SUR-
FACES. *A*, Horizontal sheeting exposed on quarry face, cut into
broad, level surface. *B*, Convex sheeting on granite dome. *C*,
Concave sheeting on valley floor. *D*, Near-vertical sheeting on
Matthes Crest. Unloading has taken place from both sides of a
steep linear ridge. Photograph from National Park Service collection. (Fig. 40)

UNDULATING SURFACE below Clouds Rest. The sheet joints are concave in the bowl-shaped basins high on the cliff face, and convex on the intervening spurs. (Fig. 41)

ROYAL ARCHES, in left center of photograph and topped by North Dome, is a gigantic expression of sheet joints, with sheets truncated by north wall of Yosemite Valley. Photograph by Eadweard Muybridge, about 1872. (Fig. 42)

WEATHERING OF JOINT BLOCKS and stages in the
formation of corestones. Corners and edges of granite blocks are
attacked more readily by weathering along joints, and rounded
corestones result. (Fig. 43)

SPHEROIDAL WEATHERING around corestones. Rec-
tangular pattern of sheets reflects horizontal and vertical orientation of joints originally bounding the disintegrating blocks.
Roadcut at Big Meadow overlook, Big Oak Flat Road.
(Fig. 44)

36

WEATHERING AND EROSION
INFLUENCE OF THE ROCKS ON WEATHERING

Unfractured granite is impermeable, and because
weathering processes depend on the presence of
moisture, exposed granite surfaces weather slowly.
However, where buried by soil and in contact with a
chemically reactive mixture of water, atmospheric
gases, and organic decay products, granite weathers
much more readily. Joints in the granite that provide
avenues for deep circulation of ground water permit
weathering to proceed well below the buried bedrock
surface. As weathering penetrates the rock from joint
surfaces, the edges and corners of the joint blocks are
affected more rapidly than the sides, because they are
attacked from two or three directions at once (fig. 43).
The unweathered remnant of granite in the center of
the joint block becomes a rounded boulder, called a
corestone, and the process of its formation is a form of
exfoliation called *spheroidal weathering* (fig. 44).

Where water collects in small natural depressions
on granitic-rock surfaces, the weathering process
commonly enlarges the depressions to form weather
pits, or pans (fig. 45). The pans are typically flat
bottomed, a fact that has not yet been completely
explained. A possible explanation is based on the
proposition that the most active environment for
weathering is the zone of alternate wetting and dry-
ing along the margins of the pools that collect in the
pans. The margins tend to deepen and enlarge until
all points of the bottom of the pan are equally wet or
dry at the same time. Thereafter, they weather downward at a rate that is constant over all of the pan
surface. This explanation also accounts for the over-
hanging rims, which are very common, and for
coalescing pans, which also are common. The flat
floors of the pans are horizontal even where the pans
occur on the sides of boulders. The weathered material in the pans is removed by wind, although in
deeper ones a granite sand remains. The process is a
slow one—such pans normally are not found on
surfaces scraped smooth during the last major glaciation, which ended some 10,000 years ago.

The weathering and disintegration of rock, making it susceptible to erosion, depend on both rock
composition and rock texture. The darker varieties of
medium-grained granitic rocks, particularly those
rich in biotite, weather more readily than the lighter
colored varieties. Expansion of the biotite by absorp-
tion of water helps free the crystal from surrounding

mineral grains and thus leads to disaggregation of the rock. The resulting granular product, a granitic sand called *grus,* is easily eroded (fig. 46). For this reason, many topographic basins in granitic terrane are in areas underlain by biotite granodiorite, and ridges are held up by granites that contain less biotite. Finer grained plutonic rocks (both light and dark colored) are generally more resistant to formation of grus than coarser grained ones. Because the fine-grained rocks occur mostly as dikes and other small bodies, they have less effect on major landforms, although they sometimes create rather spectacular features (figs. 47, 48).

The composition and texture of metamorphic rocks also affect their weathering and erosion. Again, the presence of mica facilitates breakup of the rock during weathering. The most resistant metamorphic rocks are those containing abundant quartz, such as metamorphosed sandstone, and those made dense and hard from baking by magma. Many metamorphic rocks are more resistant than plutonic rocks; metamorphic rocks hold up much of the Sierran crest along the east edge of the park, as well as the Ritter Range southeast of the park.

RESISTANT POTASSIUM FELDSPAR PHENO-CRYSTS protrude from weathered surface of Cathedral Peak Granodiorite. Matrix minerals formerly enclosing the phenocrysts have been weathered to grus (granite sand) and washed away. (Fig. 46)

KNOBS of resistant fine-grained diorite protrude from weathered outcrop surface of El Capitan Granite. (Fig. 47)

NATURAL BRIDGE formed where weather-resistant aplite bridges an opening eroded in underlying, less resistant Half Dome Granodiorite. (Fig. 48)

AGENTS OF EROSION

Erosion, simply stated, is the removal of earth materials from high areas to low areas. Erosion thus tends to level a high area.

Two agents of erosion are chiefly responsible for sculpting the present Yosemite landscape—flowing water and glacial ice. Flowing water had the major role, and glaciers added the final touches. The major drainages, the intervening divides, and the general landforms were all established before glaciation. Some of the glacial modifications were profound: the creation of alpine topography full of cirques and arêtes along the higher divides, the rounding of many valleys from V-shaped to U-shaped and their straightening in the process, and the creation of hundreds of lakes and ponds where formerly there were none.

Still another agent of erosion is simply gravity. The downslope movement of rock materials without the aid of a transport medium produces landslides and rockfalls. Although generally of local extent, such movement is important, particularly in mountainous terrain. In the winter of 1982, a rockfall dropped huge blocks of granite on Route 140 near the junction of the Old Coulterville Road, about 2 mi east of the Arch Rock Entrance Station (fig. 49). The highway remained closed until a way could be blasted through the debris, and the little-used Old Coulterville Road on the slope above was blocked severely enough to be abandoned.

ROCKFALL of blocks the size of small houses temporarily closed the El Portal Road east of Arch Rock Entrance Station in 1982 until a way could be blasted through the debris. (Fig. 49)

THE ROLE OF FLOWING WATER

The work of flowing water can be seen at all scales, from that of tiny rivulets cascading down a slope after a rain and transporting soil-size particles, to raging flood torrents with streams using stones as hammers to break up material in their beds. But flowing water can transport cobbles and boulders only during the high-energy, turbulent flow of floods. Thus, the effectiveness of erosion by flowing water depends largely on processes of weathering, the breakdown of parent rock into molecules and rock or mineral fragments that the streams can transport easily.

With the onset of late Cenozoic uplift of the range, the major streams were rejuvenated and made more vigorous by their increased gradients. In a mountain range that is rising faster than upland material can be removed, the tendency is for major streams to cut deep canyons, with both the local topographic relief and the maximum elevation of the range increasing. This canyon cutting requires that river-channel incision be faster than hillslope erosion. If river-channel incision can keep up with uplift but hillside erosion cannot, then stream channels become progressively deepened relative to areas between streams. In particular, for rocks resistant to weathering, channel incision will be relatively much faster than hillslope erosion, and a canyon is formed. Yosemite has superb examples in the canyons of the Tuolumne River, the Merced River (fig. 50), and the South Fork of the Merced River. The upper basins of these rivers were later modified by glacial erosion, but the fact that the rivers flow in deep canyons beyond the western reach of past glaciers shows that canyon cutting was accomplished solely by the action of streams.

The general courses of the major streams in Yosemite, with few exceptions, were probably inherited from the preuplift drainage pattern and depend mainly on the westerly slope of the range. The courses of the tributaries to the trunk streams and the shapes of their drainage basins depend more on granite composition and joints, as discussed in previous sections.

The results of stream incision are depicted in a series of sketches that interpret landscape development from a region of gently rolling hills with meandering streams to one of canyonlands cut into the upland surface (fig. 51). These scenes should be viewed as snapshots in a continuing process, rather than as distinct stages in landscape evolution. Note that granite domes form as the relief increases. The final scene is one conception of what Yosemite looked like about 2 or 3 million years ago, before the onset of glaciation.

DEEPLY INCISED, UNGLACIATED CANYON of the Merced River about 7 mi west of El Portal. Photograph by Dallas L. Peck. (Fig. 50)

UPLIFT AND STREAM INCISION. *A,* About 15 million years ago, the Yosemite area was a rolling surface of rounded hills and broad valleys with meandering streams. *B,* By 10 million years ago, uplift of the range was sufficient to steepen stream gradients, and the valleys deepened. *C,* Before the onset of glaciation possibly some 2 million years ago, streams had incised deep canyons into the west flank of the range. (Fig. 51)

THE ROLE OF GLACIERS

The Yosemite landscape as we see it today strongly reflects the dynamic influence of moving ice that long ago covered much of it several times. We are still uncertain as to how many times ice mantled Yosemite, but at least three major glaciations have been well documented in the Sierra Nevada, and other evidence suggests additional glacial episodes.

Formation of glaciers requires that some of the snow that falls each winter persist through the following summer into the next winter's accumulation season.

Thus, heavy winter snowfall or cool summer temperatures, or both, favor the growth of glaciers. If these conditions persist for a few centuries, possibly less, the layers of accumulating snow form a deposit thick enough for the snow in the lower part of the deposit to be compacted into ice (fig. 52). At a thickness of about 100 ft, ice begins to flow outward under its own weight, and on slopes will begin to flow downhill at lesser thicknesses; when it flows, a glacier is born.

As the glacier flows downhill, it enters regions of warmer climate, where the snow does not persist from year to year. The boundary where loss from melting and evaporation equals accumulation from snowfall is called the *annual snowline* or *firn limit*— "firn" being the term for partially compacted snow carried over from previous seasons (fig. 53). The firn limit fluctuates from year to year in response to changes in precipitation and temperature. The firn limit can be as much as 1,000 ft lower in elevation on the shaded north sides of mountain peaks than on their sunny south sides. For this reason, nearly all of the small present-day glaciers in the Sierra are on north-facing slopes.

The section of the glacier through which the maximum amount of ice flows coincides with the firn

MACLURE GLACIER, showing annual layers of ice accumulation exposed after melting of seasonal snow. Note people for scale in lower right. The layers slant upward into the glacier; the youngest layers are highest up the slope. Photograph from National Park Service collection. (Fig. 52)

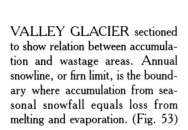

VALLEY GLACIER sectioned to show relation between accumulation and wastage areas. Annual snowline, or firn limit, is the boundary where accumulation from seasonal snowfall equals loss from melting and evaporation. (Fig. 53)

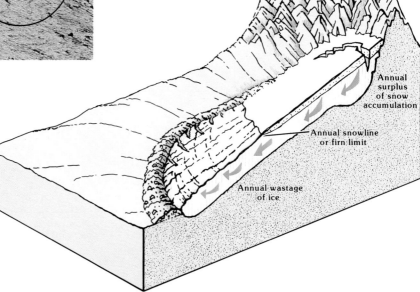

Annual surplus of snow accumulation

Annual snowline or firn limit

Annual wastage of ice

limit, because as the glacier flows toward the firn limit, it is continually augmented by new net snowfall; and downvalley from the firn limit, more ice is lost by melting and evaporation—together called *ablation*—each year than is added by snowfall. As the glacier flows downvalley from the firn limit, more and more of the ice ablates, and the glacier grows thinner or narrower, or both. Ultimately a point is reached where the ice front can advance no farther because the ice melts there as rapidly as it is provided by inflow from upglacier. If the yearly rates of accumulation and ablation were constant, this point would be fixed. However, they vary, and for that reason alone the terminus of the glacier is not likely to be fixed in position. As the climate turns warmer or drier, a glacier will gradually waste away, rather than melting catastrophically.

Glacial erosion

The velocity at which a glacier slips over its bed is only about half the ice velocity measured at its surface; the difference is taken up by deformation within the moving ice. But it is the slippage over the bed that is responsible for glacial erosion. Wherever basal flow is especially strong, or the rock easily removed, bedrock basins are carved, which eventually hold lakes. The glaciers widen and deepen valley bottoms to give a characteristic U-shaped profile. On the leesides of bedrock projections into the ice are regions of low pressure where meltwater refreezes in cracks in the rock, prying loose blocks of bedrock that are then incorporated into the glacial ice and quarried away. Asymmetric rock knobs with smoothly abraded stoss, or upstream, sides and jagged and quarried leesides are called *roches moutonnées* and record the direction of glacier flow (fig. 54). Commonly translated as "sheep," moutonnée is actually a French adjective meaning "fleecy"; the term was introduced into geology in 1786 to describe rounded Alpine hills whose repeated curves, taken as a whole and as seen from a distance, resemble a thick fleece. As rock fragments embedded in the basal ice are dragged across the bedrock surface, they impart scratches, grooves, crescentic gouges, and a shining polish (fig. 55).

Another feature of glacial erosion is the bowl-shaped, theaterlike valley head called a *cirque* (fig. 56). During the summer season, when no new snow is accumulating and the ice pulls away from the rock face at the head of the glacier, a great crevice, called a *bergschrund*, opens. This opening exposes the rock at the head of the cirque to freezing and thawing,

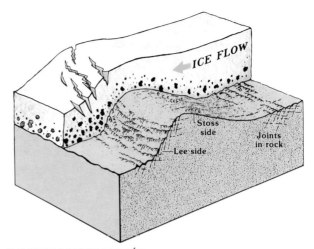

ROCHE MOUTONNÉE, sectioned to show the influence of jointing on its development. The ice moves upward over unjointed rock, smoothing it off, and plucks up and carries away blocks of jointed rock. (Fig. 54)

A

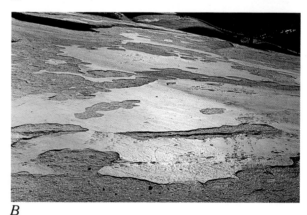

B

IMPRINTS ON THE ROCK left by passing debris-laden ice. *A*, Glacial polish and striations (lower right). Crescentic gouges, or percussion marks, are visible in center; horns of the crescents point upglacier. Chatter marks (not illustrated) consist of a group of crescent-shaped cracks pointing downglacier, but generally they do not form gouges. Photograph by John P. Lockwood. *B*, Glacial polish and striations. Polished surface layer flakes off, and this evidence of glaciation gradually disappears. This excellent and accessible exposure is at the foot of Polly Dome along the Tioga Road on the north side of Tenaya Lake. (Fig. 55)

41

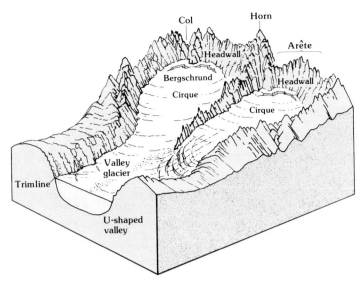

MOUNTAIN CREST, showing valley glacier and glacial sculpture. (Fig. 56)

down the cliff face. The hollows created by the detachment of these blocks and by the erosion they cause on their fall collect and hold more moisture than do the projecting ribs between them; they thus retreat more rapidly by frost wedging than do the ribs, so that even if initially smooth, a cliff face, given enough time, can become intricately sculptured.

As the cliffs on opposite sides of a ridge are quarried back by frost action above the active cirque glaciers, they eventually intersect to form a sharp, jagged rock crest called an *arête*, with sharp peaks, called *horns*, where the ridges branch (fig. 56). The sharp change in character of the walls of glaciated valleys—from intricate and jagged sculpturing above, to smooth sculpturing below—marks the edge, or *trimline*, of the former glacier and makes it possible to reconstruct the former margin of the mountain icefield (fig. 57).

Glaciated valleys are nearly straight and have U-shaped crossprofiles, in contrast to the sinuous V-shaped valleys of normal stream erosion (figs. 50, 58). At the high velocities of running water, inertia throws the fastest current against the outside of a bend. At the much slower velocities of glacial ice, the fastest flow is on the insides of bends, where the distance is shorter and the ice surface is steepest. Thus, whereas a stream erodes the outsides of bends preferentially and makes its course more sinuous, glacial erosion is concentrated on the insides of bends, removing the overlapping spurs of

which wedges blocks of rock free from the base of the cliff. The freed blocks, frozen into the main mass of ice during the winter season, are then transported out of the cirque by the glacier; retreat of the headwall thus enlarges the cirque.

The cliffs behind the cirques and above the glacier surface elsewhere are also sculpted by the freezing of water in cracks in the rock. Expansion of water when it freezes to ice breaks and wedges out blocks, which then avalanche onto the glacier's surface below. The falling blocks knock loose any projecting rock in their paths, and chip away fragments as they bounce

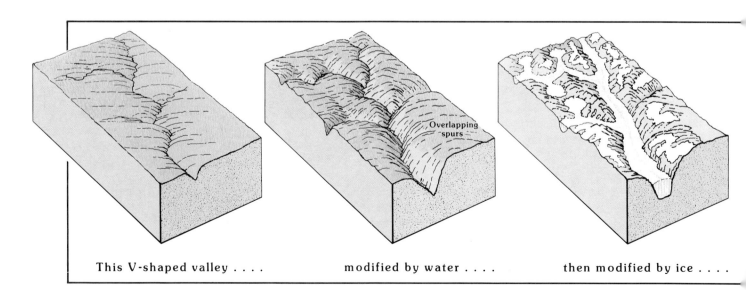

This V-shaped valley . . . .   modified by water . . . .   then modified by ice . . . .

the original stream-eroded valley and leaving a wide, straight valley floor in place of the sinuous one. A major factor leading to the U-shaped crossprofile of glacial valleys is the ability of a glacier to erode far up the valley walls. The entire glaciated valley was once occupied by the former glacier that carved it, whereas a stream occupies only the very bottom of its valley.

In addition to truncated spurs, hanging tributary valleys are formed on the sides of glaciated valleys. In a landscape developed by stream erosion, tributary streams normally join the main-trunk river at the same level. During glaciation, the upper surfaces of the glaciers occupying the tributary valleys join at the level of the surface of the ice filling the main channel; but beneath the ice the trunk glacier, with its greater thickness and erosive power, carves much more deeply into the bedrock than the tributary glaciers can. When the glacier wastes away and the ice is all gone, the tributary valleys are left hanging high up on the sides of the trunk valley, and their streams cascade or fall to join the main river below (fig. 58). During the time since the pre-Tahoe glaciation, when Yosemite Valley acquired nearly its present form, spray from the waterfalls freezing in cracks in the rock at the base of the cliff promoted spalling of rock slabs there and formed the recessed alcoves into which the falls now leap. Most of the waterfalls in Yosemite Valley formed in this manner and, indeed, are the finest examples anywhere (fig. 59). Thus, throughout the world the name "Yosemite" has come to spell "waterfalls."

JAGGED UNGLACIATED SPIRES of Unicorn Peak rising above smooth shoulders of glacially scoured granite. Boundary between jagged sculpture above and smooth below, called the trimline, marks upper limit of a former glacier along the valley wall. (Fig. 57)

VALLEY MODIFICATION by glacial erosion, showing stages in the conversion of a meandering V-shaped valley into a straightened U-shaped valley. (Fig. 58)

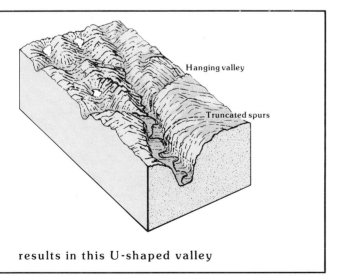

Hanging valley

Truncated spurs

results in this U-shaped valley

BRIDALVEIL FALL, an outstanding example of a waterfall issuing from a hanging valley far above Yosemite Valley floor. Photograph by Julia A. Thomas. (Fig. 59)

43

Glacial deposition

Glacially carved landforms are the most striking evidence of the passage of long-gone ice masses. But glaciers must eventually deposit the materials they are transporting, and in the process they also build up characteristic landforms. The material deposited by glaciers is an unsorted mixture of boulders, sand, and clay called *till* (fig. 60). Till may be deposited by the glacier on its bed as it is actively flowing, or the till may be left behind as the ice melts away, generally beginning as an accumulation of rock debris on the ice surface. Other depositional evidence of ice includes *glacial erratics*, boulders left behind as the ice melted (fig. 61). Direction of glacial transport is commonly indicated by boulders of rock types different from the bedrock on which they rest, but that can be traced back upglacier to a source area. Whereas till is the unsorted material deposited by a glacier, the deposit itself is known as moraine, and such moraine takes different topographic forms, depending on how the till was deposited.

The most distinctive morainal features in the Sierra Nevada are *lateral moraines*, linear ridges of till that rest on the sides of the glaciated valley and extend parallel to the valley axis (fig. 62). These ridges generally define the maximum height and width of the glacier that deposited them. Commonly, lateral moraines curve around at their lower ends to form *terminal moraines*, ridges of till deposited at the terminus of a glacier. When a glacial cycle ends, glaciers do not always melt back uniformly; instead, they commonly pause periodically in their retreat, constructing a series of *recessional moraines*. Excellent examples of all these varieties of moraines can be seen in Lee Vining Canyon just east of the park (fig. 63).

Most of the terminal and recessional moraines on the west side of the Sierra Nevada have been breached or removed by swift meltwater streams, and it is doubtful whether they were even deposited in the steep narrow canyons of most of the west-flowing streams, such as the Merced River west of Yosemite Valley. The glaciers that debouched onto the lowlands along the east base of the Sierra Nevada, however, left distinct terminal moraines, some of which presently enclose lakes.

Where two glaciers join to form a single trunk glacier, the lateral moraines being formed on their adjoining sides will continue as a linear train of debris outward onto the surface of the trunk glacier.

GLACIAL TILL, an unsorted mixture of boulders, sand, and clay, exposed along the Tioga Road at Siesta Lake. (Fig. 60)

GLACIAL ERRATIC transported by a glacier and left precariously balanced near Olmsted Point as the ice melted. (Fig. 61)

This train is called a *medial moraine*, and when the glacier melts, the debris will form a linear ridge parallel to the axis of the glacial valley. Such a medial moraine can be seen where the Merced and Tenaya Glaciers once joined at the east end of Yosemite Valley.

The total amount of till left as moraine in Yosemite and elsewhere on the west slope of the Sierra Nevada is small, however, in comparison with the amount of glacially derived debris that was flushed out of the mountains by streams swollen with glacial meltwater. Most of that debris was then deposited as alluvial fans and valley fill in the Central Valley, and some of the finer material traveled even farther, finally coming to rest in the San Francisco Bay and the Pacific Ocean.

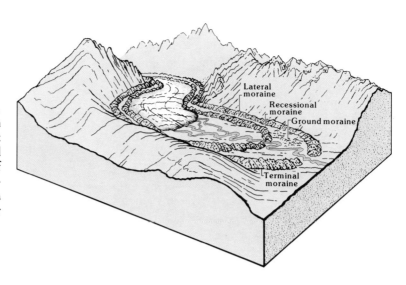

MORAINES formed by a valley glacier. Lateral moraine formed along side margin of glacier; terminal moraine formed at point of farthest glacial advance; and recessional moraine formed during pause in retreat of glacier. Ground moraine is a rather shapeless, hummocky till deposited beneath a glacier or simply left behind as the glacier retreats or wastes away. (Fig. 62)

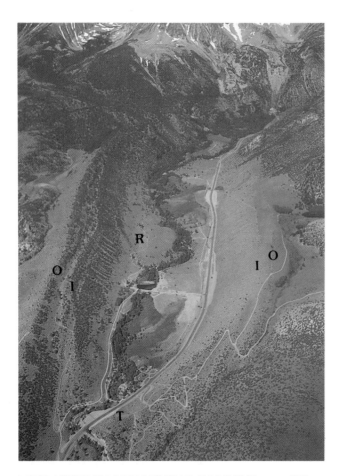

MORAINES IN LEE VINING CANYON east of Tioga Pass, looking westward toward crest of the Sierra. Paired lateral moraines on both sides of the canyon extend toward the viewer; the outer pair (O) represent the Tahoe glaciation, and the inner pair (I) the Tioga glaciation. The inner pair of moraines coalesce to form a terminal moraine (T) at lower left. A recessional moraine (R) crosses the valley as a low, tree-lined ridge. Photograph by Clyde Wahrhaftig. (Fig. 63)

THE RECORD OF PLEISTOCENE GLACIATIONS

The record of glaciation in Yosemite National Park is very incomplete. Only for the last two major glaciations can the extent of the ice be reconstructed with any confidence. Older glacial deposits, if preserved at all, are so fragmentary that it is generally impossible to distinguish the separate ice advances that may have deposited them. The glaciers grew and melted away in response to climatic changes of long duration that were probably worldwide, and so a record of all the glaciations that might have affected Yosemite must be sought elsewhere.

During worldwide glacial cycles, large volumes of water evaporated from the oceans are stored on land as ice; this storage causes a decrease in the volume of water remaining in the oceans. During evaporation of seawater, the light isotope of oxygen—oxygen-16—is more easily lost to atmospheric vapor than is the heavy isotope—oxygen-18. When this vapor precipitates on land and is stored as ice, rather than returning to the oceans, the ratio of these two isotopes of oxygen remaining in the ocean will change. Organisms living in the oceans at any given time secrete calcareous shells whose ratio of oxygen isotopes is comparable to that of the water in which the organisms lived. Measurements of fluctuations in this ratio provide indirect measurements of ice volume, with times of large ice volume interpreted as glacial episodes. Isotopic measurements on such shells extracted from deep-sea sediment samples indicate about 10 major glacial episodes during the past 1 million years. Normally, evidence for only a small fraction of these glaciations can be found in a

45

given area on land, because the more extensive glaciers destroy the moraines of earlier, less extensive ones. Only if older glaciers extended beyond the limits of younger glaciers will the older deposits remain to document the earlier glacier's existence, and so the glaciations now recognized in Yosemite may be only a fraction of those that actually occurred.

François Matthes presented evidence for three major glaciations in Yosemite, which he called, from youngest to oldest, Wisconsin, El Portal, and Glacier Point. Working on the east side of the Sierra about the same time, Eliot Blackwelder recognized four major glaciations: Tioga, Tahoe, Sherwin, and McGee; these latter terms have become firmly established and have largely replaced those of Matthes. His Wisconsin is now thought to include equivalents of both the Tioga and Tahoe glaciations, and his El Portal is probably equivalent to the Sherwin. The evidence for Matthes' separate Glacier Point glaciation is unconvincing, and there may not be an equivalent of the McGee in Yosemite. Many additional glacial episodes have been proposed to account for the diverse deposits of till, outwash, and glacial-lake sediment on the east side of the Sierra Nevada. There is disagreement, however, as to whether each of these deposits represents separate glaciations or pulses within major glaciations, and they have not been recognized in Yosemite.

In Yosemite, deposits of three separate glacial episodes are now distinguished on the following basis:

1. Tioga (youngest): Unweathered till consisting of fresh granitic boulders and loose, porous gravel and sand, making sharp-crested moraines that have closely spaced boulders on their upper surfaces (fig. 64).
2. Tahoe (intermediate): Subdued moraines whose buried granodiorite boulders have commonly disintegrated at least partly to grus. The summits of these moraines tend to be rounded, and exposed boulders along their crests are generally sparse (fig. 65).
3. Pre-Tahoe (oldest): Outside the ice limits defined by the intermediate moraines, a few glacial erratics, commonly perched (fig. 66), and patches of formless glacial till give evidence of at least one earlier ice advance more extensive than the glaciers that deposited the intermediate moraines.

By recognizing these distinctions, we can at least partly reconstruct the glacial geology of Yosemite.

THE GLACIAL GEOLOGY OF YOSEMITE

The glacially related features of today's landscape, both erosional and depositional, largely reflect the latest major glaciation, the Tioga. The effects of multiple glaciations are cumulative, however, and the effects of earlier glaciations must have been substantial, although we can only gauge them by inference. We shall examine the end results of these episodic glaciations culminating with the Tioga, and then examine the evidence for the work of earlier glaciations.

The Tioga glaciation began about 30,000 to 60,000 years ago, when a cooling climate permitted small glaciers to develop in high cirques originally formed and then abandoned by earlier glaciers. With continued cooling, these glaciers grew and moved outward and downward to coalesce into a mountain icefield, with only the higher peaks and divides projecting through as arêtes and horns. With further growth, the icefield fed fingers of ice into the major drainages on the west slope

TIOGA MORAINE in Harden Lake area, showing sharp crest and abundant boulders exposed on surface. (Fig. 64)

TAHOE MORAINE in Harden Lake area, showing subdued crest and only scattered boulders exposed on surface. (Fig. 65)

PERCHED ERRATIC deposited on ridge west of Upper Yosemite Fall during a pre-Tahoe glaciation. Pedestal height of 5 ft indicates amount of rock weathered away since boulder was dropped by the ice. Photograph from National Park Service collection. (Fig. 66)

of the Sierra, until the ice reached its maximum extent about 15,000 to 20,000 years ago (fig. 67).

The icefield in the upper Tuolumne River basin, and in the tributary basins to the north, fed the glacier that moved down the canyon of the Tuolumne River through Hetch Hetchy Valley. Some of the ice filling the basin of the Lyell Fork of the Tuolumne spilled over low passes to augment ice in the Merced River basin that flowed down through Little Yosemite Valley. Tuolumne ice also flowed over a pass into the Tenaya Lake basin and down Tenaya Canyon to join the main Merced Glacier in Yosemite Valley. During the Tioga glaciation, the glacier in Yosemite Valley reached only as far as Bridalveil Meadow.

Tioga ice also flowed eastward from the summit region to cascade down canyons cut into the east

TIOGA ICEFIELD AND VALLEY GLACIERS, showing maximum extent during the Tioga glaciation, the last major glaciation in the Sierra Nevada, which peaked about 20,000 to 15,000 years ago. (Fig. 67)

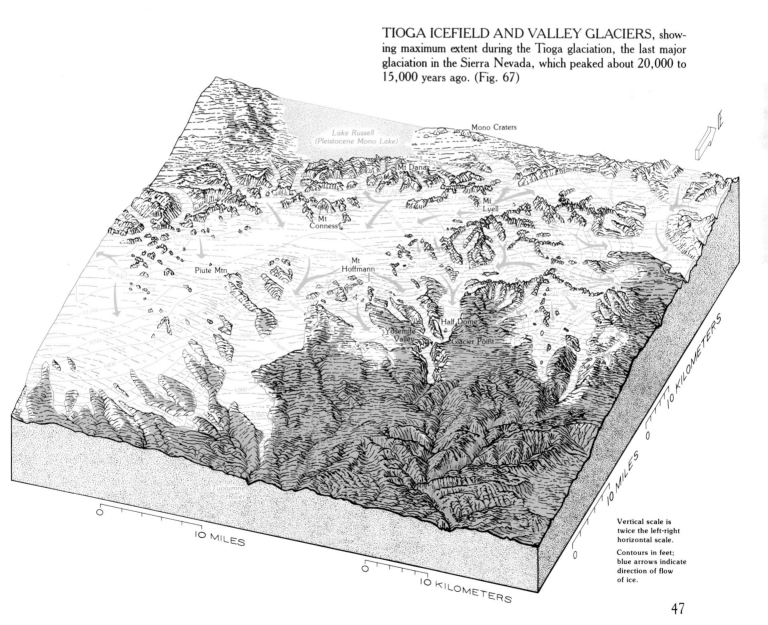

Vertical scale is twice the left-right horizontal scale.

Contours in feet; blue arrows indicate direction of flow of ice.

U-SHAPED GLACIATED VALLEY, the Lyell Fork of the Tuolumne River, looking southward. Photograph by Robert W. Cameron. © Cameron and Company; used with permission. (Fig. 68)

escarpment of the Sierra. Southeast of the park, ice from the Mount Lyell area flowed eastward onto the Mono lowland and southeastward and southward down the Middle and North Forks of the San Joaquin River. In the southern part of the park, ice in the South Fork of the Merced River reached nearly to the present site of Wawona.

In addition to the major icefields in the headwaters of the Tuolumne and Merced Rivers and the major valley glaciers they fed, smaller, isolated glaciers formed in favorable localities, as on Buena Vista Crest, Horse Ridge, and above Siesta Lake near the Tioga Road. Each played a part in creating today's landscape.

At the time of the Tioga glacial maximum, glacial Lake Russell was much larger than the present-day Mono Lake (fig. 67). The surface elevation of Lake Russell at that time was about 6,800 ft, 425 ft higher than the 1980 elevation of Mono Lake of about 6,375

ft. The increased volume of Lake Russell was probably due to a much lower rate of evaporation of water from the lake during Tioga time, partly because of the prevailing cooler climate and partly because the lake probably was covered by ice much of the time.

The Tuolumne Meadows area is one of the most accessible, as well as one of the best, places to see how glaciers modified the landscape by both erosion and deposition. The Lyell Fork of the Tuolumne River flows to the meadows through one of the park's finest examples of a U-shaped valley (fig. 68). The frost-riven spires of Unicorn and Cathedral Peaks standing above smoothly rounded granite shoulders graphically indicate the height to which Tioga ice reached (fig. 57). Glacial polish high on Fairview Dome indicates that it was overtopped by ice (fig. 69). Lembert Dome, on the northeast side of the meadows, is a roche moutonnée that records stoss-side smoothing and leeside plucking

48

on a grand scale, as does the smaller but easily ascended Pothole Dome at the meadow's west end (fig. 70). Glacial polish and striations can be seen on many outcrops, and erratic boulders abound.

A series of potholes angles diagonally from level ground up the south side of Pothole Dome (fig. 71). The rock polish near these potholes can easily be mistaken for glacial polish, but it has a different origin. Unlike glacially polished surfaces, which are generally planar, these surfaces are fluted (fig. 72), like those forming on bedrock today by flowing water of the Tuolumne River. Both the polish on this part of Pothole Dome, and the potholes themselves, were created by water in a subglacial stream that flowed upward over the dome in a tunnel beneath ice of the Tioga glacier. Glacial polish can, indeed, be seen on the gentle east-facing slope of the dome, the stoss side of this roche moutonnée.

GLACIAL POLISH high on east shoulder of Fairview Dome. It and similar polish and erratics on the summit indicate that the dome was once overtopped by ice. (Fig. 69)

POTHOLE DOME, a large roche moutonnée on the west side of Tuolumne Meadows, was shaped by glacial smoothing and plucking. Ice moved from right (stoss side) to left (leeside). Fluted surface was shaped by subglacial water scour (see fig. 72). (Fig. 70)

POTHOLES angling up side of Pothole Dome, west side of Tuolumne Meadows. Bowl-shaped potholes are carved into the rock by the grinding action of stones whirled around and kept in motion by the force of a stream in a given spot. Here, the stream flowed in a tunnel beneath the ice that covered this area during the Tioga glaciation. (Fig. 71)

SUBGLACIAL WATER POLISH on Pothole Dome, formed by water flowing beneath the glacier. Surface is fluted rather than flat, as with glacial polish (compare with fig. 55). Polished surface layer is flaking off, similar to the glacial polish shown in figure 55B. (Fig. 72)

KETTLES (small lakes) near Tioga Pass, filling depressions left by the final melting of blocks of glacial ice. View eastward across Dana Meadows toward Mount Dana. (Fig. 73)

Most of the till in the Tuolumne Meadows area was deposited as hummocky ground moraine beneath the glacier, or left behind as the glacier wasted away. The small lakes in ground moraine southeast of the entrance station at Tioga Pass appear to be *kettles*, depressions left by the final melting of blocks of ice buried in till (fig. 73). Elongate ridges of moraine flank the roadway between Tioga Pass and Tuolumne Meadows, and a cross section through such till is exposed in a roadcut about 1½ mi east of the Toulumne River bridge. Lateral and recessional moraines are inconspicuous and may be absent in the Tuolumne Meadows area, but magnificent examples of both are in lower Lee Vining Canyon along the Tioga Road east of the park (fig. 63).

Yosemite Valley, in spite of its profound glacial modification, is not a good place to see much direct evidence of glaciation. The Tioga glacier only reached as far as Bridalveil Meadow, where its terminal moraine, although well exposed in a roadcut (fig. 74), is relatively inconspicuous, and lateral moraines are absent. Glacial polish and striations are scarce, but they can be seen on a low, flat shelf of rock near the base of the north valley wall opposite the west end of the Yosemite Lodge, and along the base of the cliffs on the north side of the Tenaya Lake Trail, about 500 to 1,000 ft east of Mirror Lake. Glacial polish also occurs in such places as on the slope of the Glacier Point apron, but these localities are accessible only to the experienced climber. This paucity of direct evidence is probably the basis for Josiah Whitney's stand against the glacial origin of Yosemite Valley, noting as he did the glaciation of Hetch Hetchy Valley, where he described glacial polish at least 800 ft and a glacial moraine 1,200 ft above the valley floor.

The Tahoe glaciation was almost everywhere somewhat more extensive than the Tioga, and so its moraines lie outboard of Tioga moraines. Recent studies suggest that Yosemite Valley was an exception: a Tahoe-age glacier reached Yosemite Valley but was smaller and thinner, and did not extend as far into the valley as Tioga ice did. If both the Tioga and Tahoe glaciers had limited erosive power in Yosemite Valley because they both were relatively thin, then the valley must have attained nearly its present shape during one or more pre-Tahoe glaciations.

Direct evidence for pre-Tahoe glaciation in Yosemite Valley is also elusive. It consists mostly of scattered, commonly perched erratics on upland areas adjacent to the valley (fig. 66). These perched erratics rest on pillars of rock protected from weathering and erosion by the erratic itself. The surrounding area commonly has been lowered several feet or more by erosion since the erratics were left as the ice melted. This amount of erosion, which is not seen in areas overridden and scoured by Tahoe ice, is the main argument for attributing these erratics to a much earlier glaciation.

Readily accessible examples of pre-Tahoe erratics, though not perched, occur near the radio-relay tower on Turtleback Dome above the Wawona Road. A small patch of old, deeply weathered till is exposed in a roadcut on the Big Oak Flat Road just east of the bridge crossing Cascade Creek. This till, about 1,000 ft above the valley floor, would hardly be recognized as such without the artificial cut. These erratics and till are assigned to one or more pre-Tahoe glaciers that filled Yosemite Valley (fig. 75). No terminal moraine for this glaciation remains in the Merced River canyon, but the glacier could not have extended more than a short distance below the town of El Portal, where the canyon abruptly starts to meander and has a V shape (fig. 76).

Evidence for three separate glaciations in Yosemite is most readily seen at Harden Lake, about 3 mi down the Old Tioga Road from White Wolf. Harden Lake sits on the edge of the upland bordering the Grand Canyon of the Tuolumne River, which was completely filled with ice during each of the three recognized glaciations. Harden Lake is sandwiched between fresh, bouldery lateral moraines reflecting pulses of Tioga ice (fig. 64). Outside of these moraines, farther from the canyon, are Tahoe moraines, still-linear but subdued ridges, with extensive soil cover and only scattered boulders on the surface (fig. 65). Farthest from the canyon, outside of the Tahoe moraines, are pre-Tahoe erratics and highly dissected till whose original morainal form has been destroyed by erosion.

GLACIAL MORAINES IN YOSEMITE VALLEY. A, Tioga-age terminal moraine exposed in roadcut just east of Bridalveil Meadow. This moraine contains large boulders of Cathedral Peak Granodiorite, clear evidence of glacial transport from Tuolumne Meadows via Tenaya Canyon or from the upper Merced River basin via Little Yosemite Valley. B, Tioga-age recessional moraine exposed in roadcut below Cathedral Rock. Large boulder in center is of Cathedral Peak Granodiorite. (Fig. 74)

A

B

GLACIERS COME AND GO. Sketches of Yosemite Valley area, showing extent of pre-Tahoe glacier (A), extent of Tioga glacier (B), and glacial Lake Yosemite after retreat of Tioga ice (C). (Fig. 75)

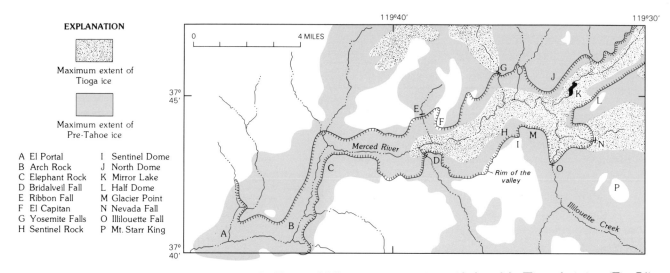

PRE-TAHOE GLACIATION—its extent in the Yosemite Valley area in comparison with that of the Tioga glaciation. (Fig. 76)

WHAT THE GLACIERS MISSED

Among the arêtes and cols of the Sierran crest and adjacent spurs are anomalous areas with gently rolling surfaces that escaped glacial sculpting; the Dana Plateau is an outstanding example (fig. 77). These areas apparently never accumulated enough snow to generate glaciers on their upland surfaces. Not only is the Sierran crest swept by strong winds, but it also receives less snowfall than areas farther west because most of the moisture is wrung out of the upwelling stormclouds before they reach the crest. Glaciers did form in cirques on the lower protected slopes of these plateaus, particularly on northerly slopes, where sufficient snow could accumulate by wind drift from the plateaus to the windward and remain through the summer. But these glaciers never succeeded in cutting cirques far enough back into the plateaus to entirely destroy them.

These upland surfaces did not totally escape the effects of glacial climate. Cycles of freezing and thawing broke the bedrock along fractures and heaved it into a mass of jumbled blocks. Such rubble-covered surfaces are scattered along the Sierran crest, as at Dana Plateau (fig. 78), and include the summit of Mount Whitney 50 mi to the south. The south-facing slope of Mount Hoffmann, our original vantage point in the park, is a somewhat similar upland surface, although slow disintegration and spalling of granite sheets has created an

DANA PLATEAU, unglaciated remnant of an ancient land surface sharply truncated by glacial cirques. Ellery Lake and the Tioga Road lie 2,000 ft below. At upper right is Mount Dana, with an unglaciated surface sloping downward to right. These two remnant surfaces are parts of a once-continuous surface breached by a glacial cirque excavated by past glaciers and now containing the present-day Dana Glacier, seen as the gray, oval-shaped spot on the snow below Mount Dana (see fig. 82). Photograph by Robert W. Cameron. © Cameron and Co.; used with permission. (Fig. 77)

BOULDER-STREWN UPLAND SURFACE on the Dana Plateau. Frost-heaved joint blocks have been rounded by sand-blasting action of wind-driven grit. Snow chute on Mount Dana near right margin leads down to the Dana Glacier across Glacier Canyon from the plateau. View southeastward. Width of view in middle distance is about ½ mi; southeast edge of plateau is about 1½ mi away. (Fig. 78)

irregular, flaggy surface. This unglaciated surface is abruptly truncated by a glacial cirque cut deeply into the north face of Hoffmann—a striking contrast in erosional processes (fig. 79).

These upland surfaces have a significance far beyond being unglaciated, because they are very ancient. They are remnants of the gently rolling terrain that existed here before the late Cenozoic uplift and incision of the Sierra that began about 25 million years ago. As the range was uplifted and tilted, the major westward-flowing streams incised deeper and deeper canyons, cutting headward into the range. The upland areas near the stream headwaters were the last to be affected—some remnants still remain undissected.

WEST SHOULDER OF MOUNT HOFFMANN. Smooth, unglaciated south-facing slope (to left) is abruptly truncated by a north-facing glacial cirque—a striking contrast in erosional processes. (Fig. 79)

MODERN GLACIERS

Although Tioga glaciers were the last to cap large parts of the Sierra Nevada, the modern Sierran glaciers are not remnants of Tioga ice. Instead, they formed during one or more post-Pleistocene cool cycles that François Matthes collectively called the "Little Ice Age." Because of ambiguity in its definition and wide variation in its application, students of glacial geology have generally abandoned the term, and post-Pleistocene glaciations are referred to as Neoglacial events. The latest of these Neoglacial events in the Sierra Nevada, to which the present-day glaciers belong, has been called the Matthes glaciation in honor of his pioneering studies.

After the end of the Pleistocene (10,000 years ago), temperatures in many parts of the world rose, reaching a maximum about 5,000 years ago. By that time, probably no glaciers survived in the Sierra Nevada. Temperatures then dropped, first slowly, then more rapidly; and small glaciers began to form in the Sierra once again by about 2,500 years ago.

World climate has continued to fluctuate through warming and cooling cycles up to the present. Historical records in the Alps indicate a major glacial advance about A.D. 1600, when glaciers descended into valleys and overwhelmed pastureland and villages. The most recent major advance there occurred about 1850. If Sierran glaciation was synchronous, Yosemite's modern glaciers would have been near their most recent maximum about the same time that John Muir was making his studies. We do know that Sierran glaciers have been receding rather rapidly since Muir's day: The first "living glacier" discovered by Muir in 1871, on Merced Peak, no longer exists.

As of 1980, there were nearly 500 glaciers remaining in the Sierra Nevada, most of them so small as to barely show evidence of iceflow, such as a bergschrund. The largest glacier left in the Sierra, the Palisade Glacier in the John Muir Wilderness, covers little more than ½ mi^2; and the largest in Yosemite, the Lyell Glacier, less than ¼ mi^2 (figs. 80, 81).

Few of the modern glaciers extended very far from their cirques, but many left rather impressive arcuate piles of moraine, considering their small size. Dana Glacier and its moraines are the most readily accessible of those remaining in the park (fig. 82); it is about a 3-hour hike upcanyon from the south shore of Tioga Lake east of Tioga Pass, but does require scrambling over large boulder piles. The glacier is now but a mere shadow of its former self as shown in a photograph taken early in the 20th century (fig. 83). This 1908 photograph shows Dana Glacier nestled up against its terminal moraine, and so it had not receded much, if any, from its latest advance. In the 67 years between the two photographs, the glacier lost about three-fourths of its area and a much larger fraction of its volume.

LYELL AND MACLURE GLACIERS. Lyell Glacier (on left) is divided into two parts by an intervening ridge. Lobate terminal moraine below indicates that at the maximum extent of the glacier, the two parts joined at the base of that ridge. Gray areas are ice exposed after melting of seasonal snow. Crevasses are especially well displayed in the ice exposed on the Maclure Glacier (on right). View southward. Photograph by Austin Post, August 1972. (Fig. 80)

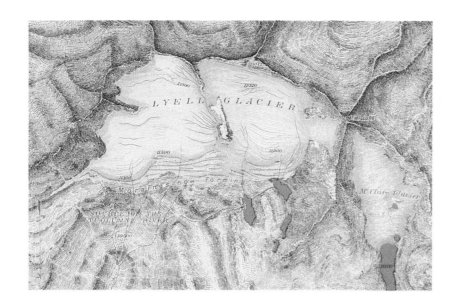

LYELL AND MACLURE GLACIERS, as they were mapped in 1883 by Willard D. Johnson of the U.S. Geological Survey. Comparison with the present-day glaciers in figure 80 indicates their much greater extent at that earlier date. Mount Maclure, named by Josiah Whitney for an early American geologist, is here mistakenly labeled "McClure." This error may be the source of occasional confusion in associating the mountain with Lt. N.F. McClure, who, however, was not on the scene as a guardian of Yosemite National Park until after 1890. (Fig. 81)

DANA GLACIER, 1975. Crevassed ice of the Dana Glacier is exposed in left center of photograph, and the glacier's bergschrund is barely visible in shadowed area above. Since the early part of the 20th century, the glacier has retreated headward from its terminal moraine, visible in lower right (compare fig. 83). View southwestward. Photograph by Malcolm M. Clark, September 1975. (Fig. 82)

DANA GLACIER, 1908. When this photograph was taken early in the 20th century, ice of the Dana Glacier abutted the terminal moraine, visible in center. Note bergschrund near glacier head below cirque headwall. View southeastward. Photograph by G.K. Gilbert, August 1908. (Fig. 83)

AFTER THE GLACIERS

The cold cycle that brought about the Tioga glaciation ended about 10,000 years ago. Because glaciers during Neoglacial episodes were so small, running water, along with gravity, has dominated further shaping of the Yosemite landscape since Tioga time. The principal changes in landscape have been the filling of lakes with rehandled Tioga till; meadows have formed as a result.

One lake to be converted to meadows was ancient Lake Yosemite, which occupied Yosemite Valley upstream from a moraine dam near the foot of Cathedral Rocks (fig. 75). This was only the last of many "Lake Yosemites" that probably followed each glaciation. The deep excavation created by earlier glaciers, as much as 2,000 ft into bedrock beneath Yosemite Valley, was aleady filled with glacial till and sediment long before Tioga time, and the Tioga ice had insufficient erosive power to reexcavate the valley to any appreciable depth. Lake Yosemite eventually filled in with silt, leaving today's level valley floor.

Gravity, commonly aided by water and ice, acting on slopes oversteepened by glacial undercutting, modifies the landscape most visibly. Canyon walls are constantly shedding rock fragments, and tremendous cones of debris accumulate in winter below avalanche chutes (fig. 84).

Rockfalls, the largest probably triggered by earthquakes, can catastrophically move large volumes of material (fig. 85). One rockfall, graphically described by John Muir, occurred in Yosemite Valley during the great Owens Valley earthquake of 1872: "The Eagle Rock on the south wall, about a half mile up the Valley, gave way and I saw it falling in thousands of the great boulders I had so long been studying, pouring to the Valley floor in a free curve luminous from friction, making a terribly sublime spectacle—an arc of glowing, passionate fire, fifteen hundred feet span, as true in form and as serene in beauty as a rainbow in the midst of the stupendous, roaring rock-storm." In May 1980, two people were critically injured on the Sierra Point Trail east of Happy Isles by rocks dislodged during a much smaller earthquake whose epicenter was near Mammoth Lakes, over 35 mi to the east. Ironically, this trail had been officially closed some years earlier, owing to the danger of loose rock. Another rockfall that killed three hikers on the Yosemite Falls Trail 6 months later may have been a delayed result of rock loosened during that same earthquake. A rockfall created the dam that formed Mirror Lake, which, in turn, is being filled by sediment carried by Tenaya Creek. Dynamic change is indeed taking place, though slowly. One lifetime is not enough to see dramatic changes.

No sooner had the glaciers departed than weathering processes began to attack the freshly scoured rock surfaces left behind. Surface spalling leaves only remnants of once-extensive glacial polish. When the polish and glacial striations are all gone, some of the most striking evidence for past glaciation will be lost.

The landscape is slowly but continually being modified. The Sierra Nevada continues to rise—and continues to be eroded. Lakes are being filled with sediment. When erosion outpaces uplift, as it eventually will, the range will be reduced to rolling upland, much as it was tens of millions of years ago. In the meantime, Yosemite remains a delight to the visitor, especially to those who learn to read the story its rocks and landscape have to tell.

AVALANCHE CHUTES AND TALUS CONES in Lee Vining Canyon east of Tioga Pass. (Fig. 84)

THE SLIDE on Slide Mountain west of Matterhorn Peak, near north boundary of the park. This giant rockslide, more than ½ mi long and ¼ mi wide, roared down with such energy that it climbed almost 200 vertical feet up the opposite side of the canyon. Photograph by Robert W. Cameron. © Cameron and Co.; used with permission. (Fig. 85)

57

DEFINITION OF TERMS

I have tried to minimize the use of technical terms in this volume, but some jargon is inevitable in any discussion of technical matters. Most of the strictly geologic terms that are apt to be stumbling blocks are defined where they first appear in the text; for ease of reference, some of them are summarized here in this short glossary. Geologic time terms are not included (see fig. 7).

ALLUVIAL FAN. A sloping, fan-shaped mass of loose rock material deposited by a stream where it emerges from a canyon onto a broad valley or plain.

ALLUVIUM. A general term for clay, silt, sand, and gravel deposited by running water, such as a stream.

ANDESITE. A volcanic rock of intermediate composition, with a silica (SiO_2) content generally of from 50 to 60 percent.

ARÊTE. A narrow, serrate mountain ridge (fig. 56).

BASALT. The most common type of volcanic rock, generally fine grained, dark, and heavy, with a silica (SiO_2) content of no more than about 50 percent.

BATHOLITH. A very large body of plutonic rock. The Sierra Nevada batholith is a composite of numerous smaller bodies (plutons) that represent repeated intrusions of granitic magma.

BEDDING. The arrangement of sedimentary rock in beds or layers, reflecting the fact that water or wind spread sediment they deposit in thin sheets. The beds of sedimentary rock are these successively accumulated sheets.

BERGSCHRUND. A deep crevice near the head of an alpine glacier that separates moving ice from the headwall of the cirque. It may be covered by or filled with snow during the winter, but visible and reopened in the summer (fig. 56).

BRECCIA. A consolidated rock composed of angular rock fragments.

CIRQUE. A bowl-shaped, theaterlike basin at the head of a glacial valley (fig. 56).

CLEAVAGE. The tendency of a mineral to break along definite planes controlled by its molecular structure and producing smooth surfaces.

COLUMNAR JOINTING. Joints that bound parallel prismatic columns, polygonal in cross section, formed by contraction during cooling in some lava flows, dikes, and volcanic plugs (a plug consists of solidified lava in an old volcanic conduit) (fig. 32).

CONGLOMERATE. A rock, the consolidated equivalent of gravel.

CRUST (OF THE EARTH). The outermost of the concentric shells that make up the Earth. The crust is 4 to 5 mi thick beneath oceans and 20 to 35 mi thick beneath continents (fig. 25).

DIKE. A sheetlike body of igneous rock that was intruded while molten into cracks in older rocks (figs. 13, 17).

DIORITE. A plutonic rock composed primarily of plagioclase and dark minerals; generally fine grained (figs. 9, 12).

ERRATIC (GLACIAL). A rock fragment, generally large, that has been transported from a distant source by the action of glacial ice (fig. 61).

EXFOLIATION. Any process by which concentric scales, plates, or shells of rock are successively spalled or stripped from the surface of a rock mass. It produces such diverse results as the spalling off of glacial polish a fraction of an inch thick and the formation of sheet joints many feet thick (figs. 39, 55).

FAULT. A fracture in the Earth's crust along which there has been movement parallel to the fracture plane.

GRADIENT. As applied to streams, the inclination of the bed.

GRANITIC ROCKS. Includes granite (in the technical sense), granodiorite, and tonalite (see fig. 9 for classification of plutonic rocks).

GRUS. The fragmental products of granular disintegration of granitic rocks in place; granitic sand (fig. 45).

HIGH SIERRA. A term coined by J.D. Whitney (1868) to describe the higher region of the Sierra Nevada, much of it above timberline.

IGNEOUS ROCK. A rock formed by solidification of hot molten material, either at depth in the Earth's crust (plutonic) or erupted at the Earth's surface (volcanic).

INTRUDE, INTRUDED, INTRUSION. Process by which magma invades or is injected into preexisting rock bodies.

INTRUSIVE SUITE. A grouping of individual plutons or plutonic rock units having significant features in common and thought to have formed from the same parent magma.

ISOTOPE. Any of two or more forms of an element with the same or very closely related properties and the same atomic number but different atomic weights. Some isotopes are radioactive and change to different isotopes at a constant rate (fig. 7).

JOINT. A fracture along which there has been little or no movement parallel to the fracture plane.

LITHOSPHERE. The rigid outer portion of the Earth comprising the Earth's crust and the uppermost part of the upper mantle (fig. 25).

MAGMA. Naturally occuring molten rock generated within the Earth. Magma may intrude to form plutonic rock or be extruded to form volcanic rock.

MANTLE (OF THE EARTH). The intermediate of the concentric shells that make up the Earth; it lies beneath the crust. (fig. 25).

METAMORPHIC ROCK. Rock changed materially in composition or appearance by heat, pressure, or infiltrations at depth in the Earth's crust.

MINERAL. A naturally occurring, inanimate substance of definite chemical composition and distinctive physical and molecular properties. Minerals make up rocks.

MORAINE. An accumulation of glacial till with an initial topographic expression of its own, commonly a ridge. Several varieties are described in the text (fig. 62).

PEGMATITE. An exceedingly coarsely crystalline plutonic rock, commonly in dikes or pods a few feet across. Individual crystals are several inches to a foot or more across.

PHENOCRYST. A large crystal in an igneous rock, embedded in a finer grained matrix.

PLATE (TECTONIC). A segment of the Earth's crust in constant motion relative to other segments (fig. 26).

PLUTON. A general term applied to any body of intrusive igneous rock of deep-seated origin, regardless of shape or size.

PLUTONIC ROCK. Igneous rock formed by solidification of magma deep within the Earth's crust.

PORPHYRITIC. Said of an igneous-rock texture with larger crystals scattered through a finer grained matrix.

PORPHYRY. A porphyritic rock with conspicuous phenocrysts in a very fine grained matrix (fig. 21).

PRE-TAHOE GLACIATION. Composite of the one or more major glaciations in Yosemite that preceded the Tahoe glaciation. Supersedes the El Portal and Glacier Point glaciations of Matthes' usage in Yosemite.

PYROCLASTIC ROCK. Rock formed of ash or other fragmental material explosively ejected from a volcano.

QUARTZ MONZONITE. A granitelike plutonic rock containing about equal proportions of potassium feldspar and plagioclase and less than 20 percent quartz under the classification system now in use. Rocks in the Yosemite area containing more than 20 percent quartz that were previously called quartz monzonite are now classified as granite (fig. 9).

RHYOLITE. A light-colored volcanic rock with a high silica (SiO_2) content of at least 70 percent.

ROCHE MOUTONNÉE. A protruding knob of bedrock glacially eroded to have a gently inclined, striated upstream slope and a steep, rough, and hackly downstream side. Commonly translated as "sheep," moutonnée is actually a French adjective meaning "fleecy"; the term was introduced into geology in 1786 in describing rounded Alpine hills whose repeated curves, taken as a whole and as seen from a distance, resemble a thick fleece (figs. 54, 70).

SCHIST. A crystalline metamorphic rock composed chiefly of platy mineral grains, such as mica, oriented so that the rock tends to split into layers or slabs.

SCHLIEREN. Streaky concentrations of dark minerals in a granitic rock, caused by movement within the partially solidified magma (fig. 16).

SEDIMENTARY ROCK. Rock formed by consolidation of sediment (gravel, sand, mud) deposited at the surface of the Earth.

SILICA (SiO_2). Occurs as the natural mineral quartz, including various fine-grained varieties, such as chert. The element silicon (Si) also occurs in most rock-forming minerals (silicates), such as feldspars.

SUBDUCTION. The process wherein an oceanic plate converging with a continental plate is deflected downward and consumed into the mantle beneath the continental plate (fig. 28).

TAHOE GLACIATION. The intermediate of the three major glaciations recognized in Yosemite. Approximately equivalent to the earlier part of the Wisconsin glaciation of Matthes.

TALUS. An accumulation of coarse, angular rock fragments derived from and resting at the base of a cliff or very steep slope.

TILL. Glacially transported material deposited directly by ice, without transportation or sorting by water (fig. 60).

TIOGA GLACIATION. The latest of the three major glaciations recognized in Yosemite. Approximately equivalent to the younger part of the Wisconsin glaciation of Matthes.

TRIMLINE. A sharp boundary line marking the maximum upper level of the margins of a glacier. It commonly separates jagged cliffs above from glacially smoothed rock surfaces below (figs. 56, 57).

TUFF. Rock formed from consolidation of volcanic ash.

REFERENCES AND ADDITIONAL READING

History

Brewer, W.H. (Farquhar, F.P., ed.), 1930, Up and down California in 1860-1864; the journal of William H. Brewer, Professor of Agriculture in the Sheffield Scientific School from 1864 to 1903: New Haven, Conn., Yale University Press, 601 p.
Fascinating account of Brewer's excursions with the California Geological Survey, including its first excursion to Yosemite in 1863.

Bunnell, L.H., 1980, Discovery of the Yosemite in 1851: Golden, Colo., Outbooks, 184 p.
Reprinted from "Discovery of the Yosemite and the Indian war of 1851 which led to that event," first published in 1880. First-hand account of the first entry into Yosemite Valley by white people.

Farquhar, F.P., 1969, History of the Sierra Nevada: Berkeley, University of California Press, 262 p.
Chapters on numerous subjects bearing on Yosemite, including the discovery of Yosemite Valley and the Big Trees by the Walker party.

Hubbard, Douglass, 1958, Ghost mines of Yosemite: P.O. Box 881, Fredericksburg, Tex., Awani Press, 32 p.
History of the Great Sierra Mine north of Tioga Pass and other prospects along the east edge of the park. "Mine" is somewhat of a misnomer because no production was ever achieved, although construction of the Great Sierra Wagon Road, forerunner of the Tioga Road, was an epic of western roadbuilding in 1883.

Hutchings, J.M., 1872, Scenes of wonder and curiosity in California; a tourist's guide to the Yosemite Valley (2d ed.): New York, A. Roman and Co., 292 p.
Hutchings, a pioneer innkeeper in Yosemite Valley, grandiloquently promoted tourism with such descriptions as that accompanying the frontispiece to this volume.

Muir, John, 1912, The Yosemite: New York, Century Co. [reprinted 1962 by Doubleday & Co., New York, Natural History Library N26, 225 p.]
Includes his graphic description of a rockfall in Yosemite Valley, triggered by the 1872 Owens Valley earthquake.

Reid, R.L., ed., 1983, A treasury of the Sierra Nevada: Berkeley, Calif., Wilderness Press, 363 p.
An excellent collection of tidbits, including Muir's description of the discovery of his first "living glacier" and the text of the Yosemite Grant to the State of California.

Russell, C.P., 1957, One hundred years in Yosemite; the story of a great park and its friends: Yosemite National Park, Calif., Yosemite Natural History Association, 195 p.
History of Yosemite from its discovery through 1956, with a detailed chronology appended. Later editions have an expanded chronology.

Whitney, J.D., 1865, Geology, volume I, Report of progress and synopsis of the field work, from 1860 to 1864: Geological Survey of California, 498 p.
Includes a description of the California Survey's first excursion to Yosemite in 1863.

Whitney, J.D., 1870, The Yosemite guide-book: Cambridge, Mass., Harvard University Press, 186 p.
Popular reprint of "The Yosemite Book" that appeared in a very limited edition in 1868 because of the inclusion of original photographic prints by Carleton E. Watkins and W. Harris. This book was the first of its kind, with the object "to call the attention of the public to the scenery of California, and to furnish a reliable guide to some of its most interesting features, namely, the Yosemite Valley, the High Sierra in its immediate vicinity, and the so-called 'Big Trees.' " That it did!

GEOLOGY

Much of the geologic history in the cited publications is outdated, except in the most recent technical ones, especially with respect to the timing of geologic events, including the uplift of the Sierra Nevada. Nevertheless, they contain much descriptive information of value and are cited in that regard.

Alpha, T.R., in press, Sketches of Yosemite National Park from Glacier Point, Sentinel Dome, and Mount Hoffmann: U.S. Geological Survey Miscellaneous Field Studies Map MF–1888 (black-and-white ed.).
Views of Yosemite physiography from three popular vantage points. Multicolored edition is published by the Yosemite Institute.

Balogh, Richard, 1977, Where water flowed "uphill" on Pothole Dome: Yosemite National Park, Calif., Yosemite Natural History Association, Yosemite Nature Notes, v. 46, no. 2, p. 34–39.
Description of the effects of subglacial stream action on Pothole Dome.

Bateman, P.C., 1983, A summary of critical relations in the central part of the Sierra Nevada batholith, California, U.S.A., in Roddick, J.A., ed., Circum-Pacific plutonic terranes: Geological Society of America Memoir 159, p. 241–254.
The most comprehensive technical summary of the geology of the Sierra Nevada batholith to date.

Bateman, P.C., in press a, Constitution and genesis of the central part of the Sierra Nevada batholith: U.S. Geological Survey Professional Paper.
A technical report that summarizes the rationale for separating and naming individual plutonic-rock units and intrusive suites in the central Sierra Nevada, including the Yosemite area. This report supersedes that by Bateman (1983).

Bateman, P.C., and Chappell, B.W., 1979, Crystallization, fractionation, and solidification of the Tuolumne Intrusive Series, Yosemite National Park, California: Geological Society America of Bulletin, pt. 1, v. 90, p. 465–482.

Technical analysis of the geology of the Tuolumne Intrusive Suite (formerly called the "Tuolumne Intrusive Series").

Bateman, P.C., and Wahrhaftig, Clyde, 1966, Geology of the Sierra Nevada, *in* Bailey, E.H., ed., Geology of northern California: California Division of Mines and Geology Bulletin 190, p. 107–172.

Though written before the advent of the concept of plate tectonics, this article remains the best general descriptive summary of the geology of the Sierra Nevada.

Cameron, Robert, 1983, Above Yosemite, *with a text by* Harold Gilliam: San Francisco, Cameron and Co., 144 p.

Superb collection of color photographs of Yosemite as seen from the air, many illustrating geologic features as well as beautiful scenery.

Fryxell, Fritiof, ed., 1962, François Matthes and the marks of time; Yosemite and the High Sierra: San Francisco, Sierra Club, 189 p.

Collection of geological essays by François Matthes, written for a lay audience.

Hill, Mary, 1975, Geology of the Sierra Nevada: Berkeley, University of California Press, 232 p.

Geologic introduction to the Sierra Nevada in nontechnical language.

Hill, Mary, 1984, California landscape, origin and development: Berkeley, University of California Press, 262 p.

Describes in nontechnical language the geologic processes that contribute to landscape development, with many examples selected from the Sierra Nevada.

Huber, N.K., 1981, Amount and timing of late Cenozoic uplift and tilt of the central Sierra Nevada—evidence from the upper San Joaquin River basin: U.S. Geological Survey Professional Paper 1197, 28 p.

Concludes that the late Cenozoic uplift was relatively continuous rather than episodic and that the current uplift rate at Deadman Pass west of Mammoth Lakes is about 1½ inches per 100 years, a figure that in this volume has been extrapolated to Mount Dana.

Jones, W.R., 1976, Domes, cliffs, and waterfalls; a brief geology of Yosemite Valley: Yosemite National Park, Calif., Yosemite Natural History Association, 21 p.

Nicely illustrated booklet, with brief geologic summary.

Jones, W.R., 1981, Ten trail trips in Yosemite National Park: Golden, Colo., Outbooks, 78 p.

Includes a description of a hike up Mount Hoffmann, with comments on the geology along the way.

Jones, W.R., 1981, Yosemite, the story behind the scenery (revised ed.): Las Vegas, Nev., KC Publications, 48 p.

Beautifully illustrated book on geologic and other natural-history subjects.

Matthes, F.E., 1930, Geologic history of the Yosemite Valley, *with an appendix on* The granitic rocks of the Yosemite region, by F.C. Calkins: U.S. Geological Survey Professional Paper 160, 137 p.

A classic study in glacial geology. Some of its conclusions have been challenged, but it remains the best source of descriptive material on the glaciation of Yosemite.

Matthes, F.E. (Fryxell, Fritiof, ed.), 1950, The incomparable valley; a geologic interpretation of the Yosemite: Berkeley, University of California Press, 160 p.

An eloquent version of Matthes' classic study, expressly written for the lay reader.

Osborne, Michael, 1983, Granite, water and light; waterfalls of Yosemite Valley: Yosemite National Park, Calif., Yosemite Natural History Association, 48 p.

Outstanding photographs, focusing on the waterfalls. Brief geology.

Russell, I.C., 1889, Quaternary history of Mono Valley, California: U.S. Geological Survey Annual Report 8, p. 261–394 [reprinted 1984 by Artemisia Press, P.O. Box 119, Lee Vining, CA 93541].

Russell blends science with prose so vivid that this report is essential reading for anyone with an interest in the Mono Basin, Mono Craters, and the glaciers of the adjacent High Sierra, both present and past. Illustrated with etchings and the superb cartography of Willard D. Johnson (see fig. 81 of this volume). Glacial Lake Russell, which occupied the Mono Basin during the Tioga Glaciation, was so named in recognition of this pioneering work.

Schaffer, J.P., 1983, Yosemite National Park—a natural-history guide to Yosemite and its trails (2d ed.): Berkeley, Calif., Wilderness Press, 274 p.

An excellent trailguide, with brief commentaries on the geologic features along many routes.

Sharp, R.P., 1960, Glaciers: Eugene, Oregon System of Higher Education, Condon Lecture Publications, 78 p.

An illuminating discussion of glaciers for the nonspecialist.

GEOLOGIC MAPS

These maps show details of geology not possible to include in this volume, and together with yet unpublished maps they form the foundation upon which I have constructed the geologic story.

Alpha, T.R., Huber, N.K., and Wahrhaftig, Clyde, 1986, Oblique map of Yosemite National Park, California: U.S. Geological Survey Miscellaneous Investigations Series Map I–1776.

This map depicts Yosemite's landforms as viewed looking down toward the northeast at an angle of 30°. It is an enlarged and more detailed version of figure 6 in this volume.

Alpha, T.R., Wahrhaftig, Clyde, and Huber, N.K., 1987, Oblique map showing the maximum extent of 20,000-year-old (Tioga) glaciers, Yosemite National Park, California: U.S. Geological Survey Miscellaneous Investigations Series Map I–1885.

This map depicts the Tioga icefield and associated valley glaciers as viewed looking down toward the northeast at an angle of 30°. It is an enlarged and more detailed version of figure 67 in this volume.

Bateman, P.C., in press b, Pre-Tertiary bedrock geology of the Mariposa 1° × 2° quadrangle, central Sierra Nevada, California: U.S. Geological Survey Miscellaneous Investigations Series Map.

This geologic map shows the bedrock geology with Cenozoic volcanic rocks and surficial deposits stripped off so as to concentrate on the plutonic rocks of the Sierra Nevada batholith and associated metamorphic rocks. All but the north one-fourth of Yosemite National Park lies within the limits of this map.

Bateman, P.C., in press c, Geologic map of the Bass Lake quadrangle, west-central Sierra Nevada, California: U.S. Geological Survey Geologic Quadrangle Map.

Bateman, P.C., Kistler, R.W., Peck, D.L., and Busacca, A.J., 1983, Geologic map of the Tuolomne Meadows quadrangle, Yosemite National Park, California: U.S. Geological Survey Geologic Quadrangle Map GQ–1570, scale 1:62,500.

Bateman, P.C., and Krauskopf, K.B., 1987, Geologic map of the El Portal quadrangle, west-central Sierra Nevada, California: U.S. Geological Survey Geologic Quadrangle Map.

Calkins, F.C., 1985, Bedrock geologic map of Yosemite Valley, Yosemite National Park, California, *with accompanying pamphlet by* N.K. Huber and J.A. Roller: U.S. Geological Survey Miscellaneous Investigations Series Map I-1639, scale 1:24,000.

Chesterman, C.W., 1975, Geology of the Matterhorn Peak 15-minute quadrangle, Mono and Tuolumne Counties, California: California Division of Mines and Geology Map Sheet 22, scale 1:48,000.

Chesterman, C.W., and Gray, C.H., Jr., 1975, Geology of the Bodie 15-minute quadrangle, Mono County, California: California Division of Mines and Geology Map Sheet 21, scale 1:48,000.

Huber, N.K., 1983, Preliminary geologic map of the Pinecrest quadrangle, central Sierra Nevada, California: U.S. Geological Survey Miscellaneous Field Studies Map MF–1437, scale 1:62,500.

Huber, N.K., Bateman, P.C., and Wahrhaftig, Clyde, compilers, in press, Geologic map of Yosemite National Park and vicinity, Calilfornia: U.S. Geological Survey Miscellaneous Investigations Series Map I–1874, scale 1:125,000.

Huber, N.K., and Rinehart, C.D., 1965, Geologic map of the Devils Postpile quadrangle, Sierra Nevada, California: U.S. Geological Survey Geologic Quadrangle Map GQ–437, scale 1:62,500.

Keith, W.J., and Seitz, J.F., 1981, Geologic map of the Hoover Wilderness and adjacent study area, Mono and Tuolumne Counties, California: U.S. Geological Survey Miscellaneous Field Studies Map MF–1101–A, scale 1:62,500.

Kistler, R.W., 1966, Geologic map of the Mono Craters quadrangle, Mono and Tuolumne Counties, California: U.S. Geological Survey Geologic Quadrangle Map GQ–462, scale 1:62,500.

Kistler, R.W., 1973, Geologic map of the Hetch Hetchy Reservoir quadrangle, Yosemite National Park, California: U.S. Geological Survey Geologic Quadrangle Map GQ–1112, scale 1:62,500.

Peck, D.L., 1980, Geologic map of the Merced Peak quadrangle, central Sierra Nevada, California: U.S. Geological Survey Geologic Quadrangle Map GQ–1531, scale 1:62,500.

Wahrhaftig, Clyde, in press, Glacial map of Yosemite National Park and vicinity, California: U.S. Geological Survey Miscellaneous Investigations Series Map.

Map showing the maximum extent of Tioga-age glaciers and the distribution of glacial moraines of various ages. It provides the technical background for figure 67 in this volume (Tioga icefield and valley glaciers).

A WORD OF THANKS

Geologist Alfred C. Lane once wrote, "The progress of knowledge is like the growth of a coral reef; each generation builds upon that which has been left behind by those who have gone before." So it is with this volume. I have drawn upon so many sources of information in presenting this geologic story of Yosemite that it is impossible to acknowledge the individual contributions of each. This is especially true of the historical material and of the numerous works that have delineated the regional geologic framework within which I have placed Yosemite. For those who wish to delve further into various aspects of Yosemite history and geology, several pertinent references, briefly annotated, are listed in the bibliography; these, in turn, will lead to additional source materials.

I must, however, specifically acknowledge the modern geologic mapping and detailed studies by my U.S. Geological Survey colleagues that made possible the present volume, as well as the new geologic map of Yosemite National Park published separately (Huber and others, in press). For that foundation and for their continuing support and contributions, I thank Paul Bateman, Lew Calk, Frank Dodge, Bill Keith, Ron Kistler, Dallas Peck, Dean Rinehart, Jim Seitz, and Clyde Wahrhaftig. Clyde Wahrhaftig also developed the explanation for the formation of weather pans, described on page 36.

Julie Roller assisted me in the field and in the early stages of writing this volume and compiling the geologic map. Thoughtful manuscript reviews were provided by many of my colleagues, as well as by Genny Smith of Palo Alto and Mammoth Lakes, Calif., and Jim Sano of the National Park Service. Finally, during excursions to Yosemite, the hospitality and enthusiastic support of every member of the Park Service staff there contributed significantly to the success of the project. Jan van Wagtendonk was particularly helpful in assisting with logistic support.

Thanks are also due to the individuals and organizations permitting use of their photographs; those not by myself are credited in the captions, except for the frontispiece, which was contributed by Dallas Peck. Throughout the process of illustrating this volume, I have had the distinct pleasure of drawing upon and working with the artistic talents of cartographer Tau Rho Alpha and illustrator Susan Mayfield. In particular, Tau created the oblique views of Yosemite's physiography (fig. 6) and the Tioga glaciation (fig. 67), and the panorama from Mount Hoffman (fig. 5). The oblique views (figs. 6, 67) have been published in much-enlarged, more detailed versions as U.S. Geological Survey Miscellaneous Investigations Series Maps I–1776 and I–1885, respectively.

INDEX

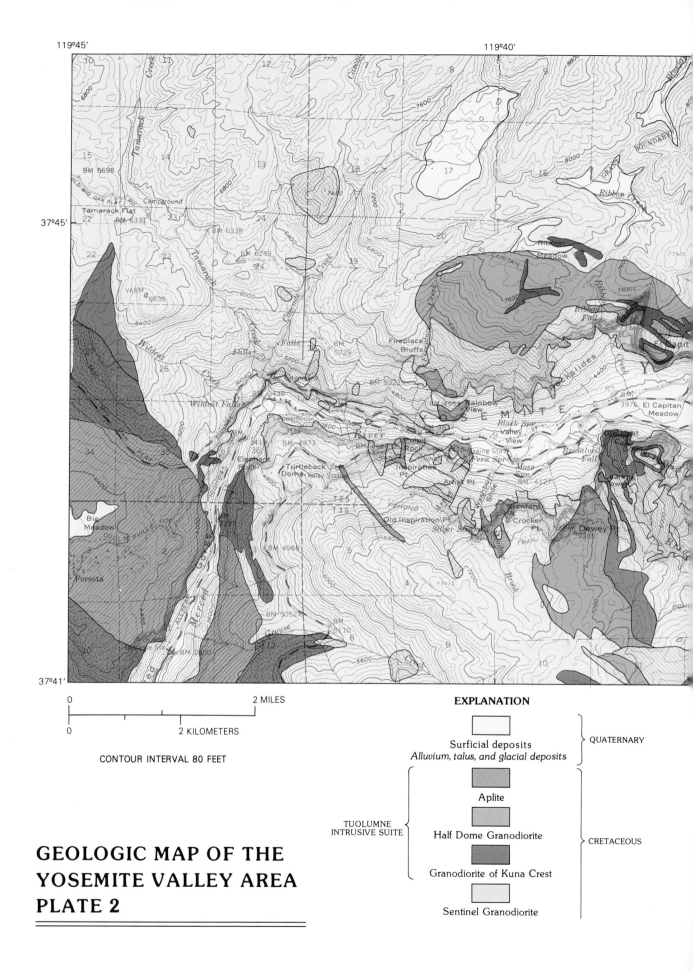

GEOLOGIC MAP OF THE YOSEMITE VALLEY AREA
PLATE 2

0   2 MILES

0   2 KILOMETERS

CONTOUR INTERVAL 80 FEET

EXPLANATION

Surficial deposits
Alluvium, talus, and glacial deposits   QUATERNARY

Aplite

TUOLUMNE
INTRUSIVE SUITE

Half Dome Granodiorite   CRETACEOUS

Granodiorite of Kuna Crest

Sentinel Granodiorite

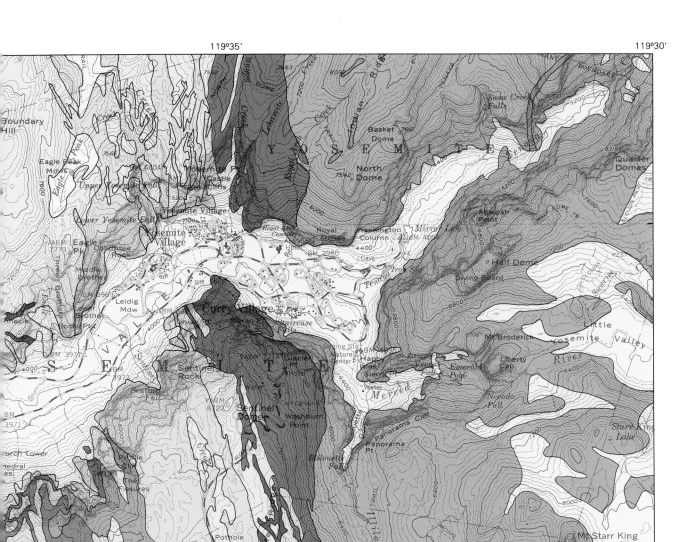

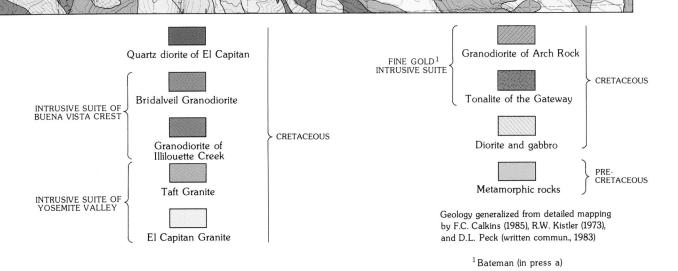

Quartz diorite of El Capitan

INTRUSIVE SUITE OF
BUENA VISTA CREST

Bridalveil Granodiorite

Granodiorite of
Illilouette Creek

CRETACEOUS

INTRUSIVE SUITE OF
YOSEMITE VALLEY

Taft Granite

El Capitan Granite

FINE GOLD[1]
INTRUSIVE SUITE

Granodiorite of Arch Rock

Tonalite of the Gateway

CRETACEOUS

Diorite and gabbro

Metamorphic rocks

PRE-
CRETACEOUS

Geology generalized from detailed mapping
by F.C. Calkins (1985), R.W. Kistler (1973),
and D.L. Peck (written commun., 1983)

[1] Bateman (in press a)

The Geologic Story of
YOSEMITE
NATIONAL PARK